# Africa and Climate Change

Ibukun Olukorede Popoola

# Africa and Climate Change

**Ibukun Olukorede Popoola**

**Academica Press**
**Washington~London**

Library of Congress Cataloging-in-Publication Data

Names: Popoola, Ibukun Olukorede (author)
Title: Africa and climate change | Popoola, Ibukun Olukorede.
Description: Washington : Academica Press, 2022. | Includes references.
Identifiers: LCCN 2022944784 | ISBN 9781680537956 (hardcover) | 9781680537963 (e-book)

**This book is dedicated**

To

*The Sovereign God, the Excellency of the Supreme body of eternal knowledge*

*My Beloved Parents, whose anchor of encouragement I grip*

*My Sweetheart and Research partner, whose supports remain in my chronicles*

*My Esteemed Supervisor, who left me with a seed of scientific virtues*

*My Beloved Son and Siblings, whose contributions gladden my heart.*

# Acknowledgements

The efforts of the following individuals deserve special mention

Engr. M. O. A. Popoola and Mrs Abigael Popoola, Director of Graadseko Construction Ltd, for full sponsorship; Oluwabukola Rashidat Popoola, Research Scientist at Agricultural Research and Biometrics Department, Product and Processing Unit, Thisavrous Pyrgos Integrated Resources, Nigeria; Professor T.O. Ososanya, Department of Animal Science, University of Ibadan, Nigeria; Dr O. A. Abu, Department of Animal Science, University of Ibadan, Nigeria; Dr. O. A. Adebiyi, Department of Animal Science, University of Ibadan, Nigeria; Dr. O. Odu, Department of Animal Science, University of Ibadan, Nigeria; Olusegun and Oludele Popoola (Director at Excellent Achievers Concept, Nigeria); Aladeniyi Timileyin, Manager at Timi Printing press; Salau Taofeek, Scientist at Central Laboratory, Department of Animal Science, University of Ibadan, Nigeria; Dr. Deborah Damilola Popoola; Nancy Ho, Managing Editor at Scientific Research Publishing, USA; Jessica Hu, Editor at Open Journal of Animal Science, SRP, USA; Evang. Cecilia Sijuade Adeola, Founder of Jesus Power Ministries, Nigeria; Ezekiel Akinbiyi Mark, General Overseer, Christ Power Evangelical crusade International, for ceaseless covenant prayers of faith.

# Contents

**Acknowledgements ........ vii**

**Chapter One**
**The Uniqueness of Africa: Why Africa? ........ 1**

1.0. The Uniqueness of Africa: Why Africa? ........ 1

1.1 Geographical evaluation of Africa ........ 2

1.1.1. Features of Africa`s physical regions ........ 2

(a) Sahara ........ 2

(b) Rain forest ........ 3

(c) Sahel ........ 3

(d) Savanna ........ 3

(e) Swahili coast ........ 3

(f) Ethiopian highlands ........ 4

(g) Southern Africa ........ 4

(h) African Great Lakes ........ 4

1.2. Africa`s Vulnerable Status in Global Climate Change ........ 6

**Chapter Two**
**The Degradability of the Life-enhancing Planet- Earth ........ 9**

2.0. The Degradability of the Life-enhancing Planet- Earth ........ 9

2.1. The Earth`s Concentric Layers ........ 9

**Chapter Three**
**Living Organisms Predisposed to Changes in Global Climate ........ 13**

3.0. Living Organisms Predisposed to Changes in Global Climate ....... 13

3.1. Plants ........................................................................................ 14

3.2. Micro-organisms.......................................................................... 19

3.2.1. Disease spread in microbial community........................................ 30

3.3. Humans and ecosystem networks.................................................. 33

**Chapter Four**
**Adaptation Measures for Changes in Global Climate ...................... 43**

4.0. Adaptation Measures for Changes in Global Climate..................... 43

4.1. Microbial adaptation to changes in global climate.......................... 44

4.2. The need for adaptation ................................................................ 46

4.3. The extent of adaptation strategies ................................................ 48

4.4. Processes of adaptation................................................................. 49

4.5. Partners currently involved
in adaptation and mitigation in vulnerable regions ............................... 52

**Bibliography........................................................................................ 55**

# Figures


| | | |
|---|---|---|
| **Figure 1:** | **The Uniqueness and Vulnerability of Africa** | **5** |
| **Figure 2:** | **Effects of climate change on living organisms** | **14** |
| **Figure 3:** | **Adaptation strategies to ameliorate the adverse effects of climate change on microorganisms, plants, humans and animals** | **44** |

# Chapter One

# The Uniqueness of Africa: Why Africa?

## 1.0. The Uniqueness of Africa: Why Africa?

Africa, though a continent, has been referred to as the home to the most genetically diverse people on Earth, with an estimated number of over 1.2 billion inhabitants that accounts for about 17% of the world`s population as at the year 2018. It is the second largest continent and accounts for about 20% of the total land surface of the earth. Africa`s estimated total area is about 30.37 million square kilometers and covers about 5.7% of the earth's surface uniformly, with assumed equal vertical and horizontal dimensions of 7,248km and 7,009km, respectively. Africa has approximately 30% of the earth's remaining mineral resources, and has the largest reserves of precious metals with over 90% of the platinum reserves, 40% of the gold reserves, over 60% of the cobalt.

The continent has five cardinally named sub sections (North Africa, Southern Africa, East Africa, West Africa, and Central Africa) with their unique characteristics. Fossil report revealed age-long evidences that Africa seems to be the ancient continent where humans were found and there are at least over 3,000 ethnic groups with about 2,000 spoken languages. Also, the world`s fastest (cheetah), tallest (giraffes) and largest (African elephant) animals are found in Africa and it is a home (Tanzania) to the largest wildlife immigration on earth. Africa has over 25% of the world's bird species and over 85% of the world's elephants, while over 99% of the remaining lions are on the continent.

The world`s longest river- Nile has its source from Africa, with an estimated length of 6,853 km that cut across eleven countries in Africa. Africa is also unique for having the World`s largest hottest desert (Sahara desert) that is referred to as the harshest environment on earth and is bigger than USA, with an estimated 9.4 million square kilometres. It is also the world`s third largest desert, after Antarctica and Arctic which are cold deserts. This uniqueness might also predispose the mother continent to severe heat stress conditions (Norman, 2018).

## 1.1 Geographical evaluation of Africa

Africa is bounded by the Mediterranean Sea, the Red Sea, the Indian Ocean, and the Atlantic Ocean and characterized by diverse ecological systems. It is divided in half almost equally by the Equator and has 8 major physical regions that includes the Sahara, Rain forest, Sahel, Savanna, Swahili coast, Ethiopian highlands, Southern Africa and African Great Lakes with unique animal and plant communities.

### 1.1.1. Features of Africa`s physical regions

#### (a) Sahara

The Sahara desert is the world`s largest hot desert and makes up about 25% of the continent. It is characterized by (i) regs which makes up about 70% of the Sahara and are plains of sand and gravel; (ii) ergs which covers about 20% of the desert and contains large quantities of salts with industrial benefits; (iii) hamadas, which are elevated plateaus of rocks and stones that reach heights of 3,353 meters; and (iv) oases which is a hub of water in the desert, and is the major source of water for the Saharan population often in the form of springs, wells, or irrigation systems.

For plants and animals resident in the Sahara, distinct adaptive features have been exhibited which allow them cope with the extreme dry conditions of the region. Documented examples include the plunging nature of plants roots extending as far as 24 metres underground. Animals such as dromedary camel retain water by changing its body temperature to accommodate the heat load, Kidneys of the jerboa produces highly concentrated urine being a rodent and minimizes water loss to compensate for heat stress, while some limits their metabolic and physical activities to the night when temperature declines. Others absorb water from their prey and adjust to extreme environmental temperatures.

## (b) Rain forest

The rain forests of Africa have diverse varieties of plants and animals with unique properties. For example, the driver ants can move across the rain forest floor, without any preference for food types and may consume anything nature sets from toxic millipedes to reptiles and small mammals. However, most of the native rain forests have been destroyed by human activities.

## (c) Sahel

The Sahel is made up of plains of flat and barren properties and stretches across Africa at about 5,400 kilometers from Senegal to Sudan. It is narrow, and forms a transition between the Sahara and the Savanna. A unique feature of the Sahel is that it contains the fertile delta of the Niger, which is one of Africa`s longest rivers. Unfortunately, these fertile lands are rapidly becoming deserted as a result of drought, deforestation, and intensive agriculture. Despite these uniqueness, the Sahelian animal communities are constantly scavenging for water and vegetation.

## (d) Savanna

Savanna covers about 50% of Africa, with the Serengeti plains mostly recognized for the vast, undulating structure among the savanna regions and as the home to most large mammal species. A continuous annual circular migration of more than a million animals occur across the Serengeti plains and exert a symbiotic relationship of soil fertilization and regrowth of new plant species.

## (e) Swahili coast

One interesting fact about the Swahili coast is that there are limited animal types residing in the region when compared to the Savanna. The coast extends about 1,610km along the Indian Ocean and is being protected by reefs and barrier islands from severe weather conditions. Human activities in the interior areas of the coast have diminished plant species growth as a result of agriculture. The most common vegetation is

Mangrove forests due to their exposed root systems that permit oxygen absorption from the air and from soils.

### (f) Ethiopian highlands

The Ethiopian Highlands are home to most of Africa`s tallest mountains and began as magma from earth`s mantle in ancient times, and was later split to Africa`s continental crust, creating the Great Rift Valley system. The highlands craggy landscape is perfect for nimble animal species and the most emblematic highlands species is probably the Ethiopian wolf, which is now on the brink of extinction.

### (g) Southern Africa

Southern Africa is dominated by a shelf of bedrock from ancient times and includes rocky features such as plateaus and mountains. It is referred to as the epicenter of Africa`s well-known reserves, which protect animal species such as elephants, lions, baboons, white rhinos, and Burchells zebras. Other important animal species include the impala, a type of deer, and the springbok, a type of gazelle that can spring several feet into the air to avoid predators. Also, floral region is one of the richest plant areas in the world.

### (h) African Great Lakes

The geologic processes of deep cracks created in the earth`s surface have led to the creation of some of the largest and deepest lakes in the world. There are seven major African Great Lakes and includes Lake Albert, Lake Edward, Lake Kivu, Lake Malawi, Lake Tanganyika, Lake Turkana, and Lake Victoria which is the largest lake in Africa, and known to be the southern source of the longest river in the world (Nile). This Great Lake has diverse range of aquatic and terrestrial animal life.

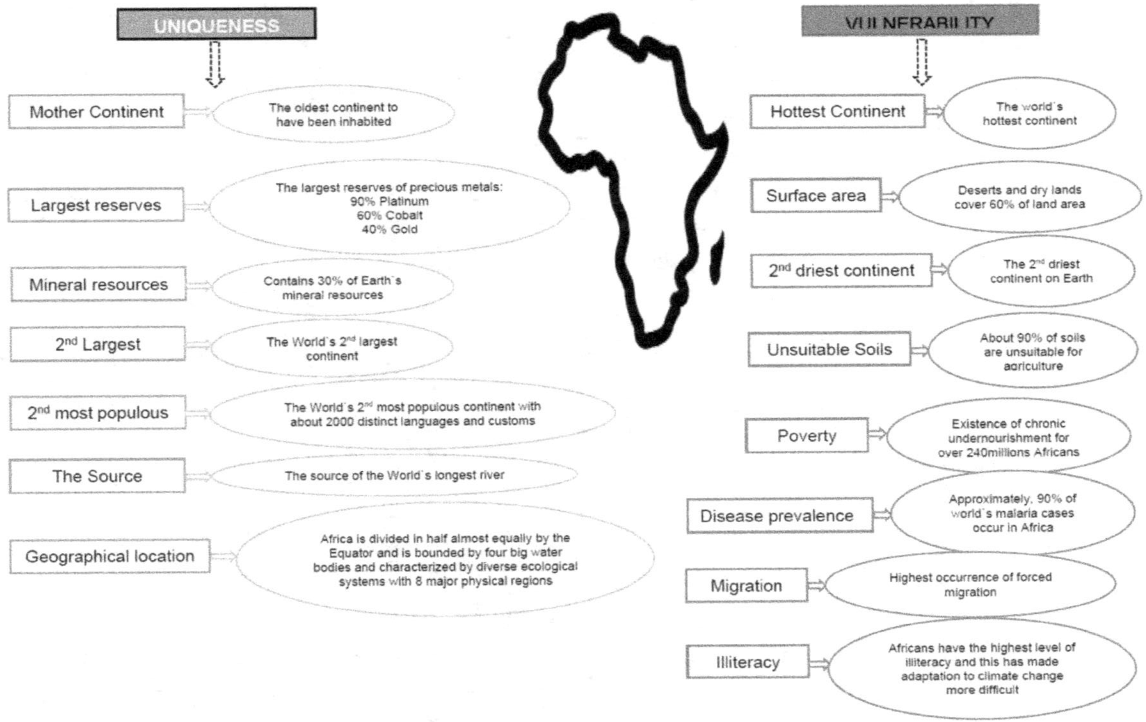


**Figure 1: The uniqueness and vulnerability of Africa**

## 1.2. Africa`s Vulnerable Status in Global Climate Change

Africa is the world's hottest continent with deserts and dry lands covering about 60% of land surface area. This attribute has positioned Africa has the most vulnerable continent to changes in the Earth`s surface temperature. It is also the world's second driest continent after Australia. Africa records about 90% of malaria cases and stands as the highest continent with occurrence of malaria cases, with about 24% of child death in sub-Saharan Africa.

The illiteracy level of Africans cannot be overlooked, as almost 40% of African adults are illiterates, with about two-thirds being women. In certain countries, literacy rates are ranked below average in adults, and there are fewer people with internet connections in Africa than there are in New York City alone. Despite the fact that a higher percentage of Africa`s work force are found in food production and efficient utilization of vast areas of arable and pastoral lands, over 230 million Africans suffer from chronic undernourishment. One reason for this could have been as a result of over 90% of Africa`s soils being unsuitable for agricultural purposes. Rainfall variability is very high in Africa and ranges from 0 mm/year in the Sahara to about 9,500 mm/year in Cameroon. This variability engineered by global warming has allowed for severe scarcity of water, affecting over 300 million Africans. Agricultural productivity has declined due to eroded soil structure and poor land management practices, with some landscapes loosing over 50 metrics tonnes of soil per hectare in a year. Although, some African countries such as Nigeria, Algeria, Angola, Libya, Egypt, Sudan, Equatorial Guinea, Republic of Congo, Gabon, and South Africa are rich in crude oil, yet unmanaged pollutions over the years have limited agricultural success. Africa has been labelled as having the highest rate of deforestation, which is twice the average for the rest of the world. What will then replace the covering potentials of these forest trees against the changing climate, when over 20% of land surface is being exposed? Even though the exploitation of these forest trees are considered commercially beneficial, yet increasing illiteracy level and improper sensitization about climate change have further triggered these unfavorable events. The most extended biomass burning in the world occur in Africa and yet only emits about 4% of the world's total carbon dioxide emissions (Norman, 2018).

# Chapter Two

# The Degradability of the Life-enhancing Planet- Earth

## 2.0. The Degradability of the Life-enhancing Planet-Earth

Complex interactions occur among the Sun, atmosphere, cryosphere, biosphere, oceans, and land surface based on the laws of conservation of mass and energy. The principal force driver is the Sun and determines the weather and climate pattern by unevenly distributing its energy on the Earth`s surface due to tilting in the earth`s axis of rotation. Within a year, the angle of rotation results in equatorial areas receiving more solar energy than those near the poles. As a result, the tropical oceans and land masses absorb a greater heat than other regions of Earth, while the atmosphere and oceans act together to redistribute the heat. As the equatorial waters warm air near the ocean surface, it expands and carries heat and moisture towards the poles and cooler denser air from the subtropics and the poles moves toward the equator to take its place. This continuous heat distribution is modified by the west to east rotation of the planet and the Coriolis force associated with the spherical shape of the Earth, and consequently result in high jet streams and winds that drive ocean currents in a pattern that moderates climate change periodically (Monroe and Wicander, 1994).

## 2.1. The Earth`s Concentric Layers

The core, mantle and crust are the earth`s three concentric layers with distinct chemical and physical properties. The core has solid and liquid layers. The solid layer is found beneath the outer core that is liquid. The core is mostly iron, with some nickel and takes up 16% of Earth's total volume. The metallic core accounts for Earth's magnetic field through the joint action of the liquid metal within the outer core and the Earth's rotation. However, the intensity of magnetism may change from time to time within decades and over the period of 500 000 years for complete reversal as evidenced in the rocks. The strength of the earth can be determined by the critical evaluation of some rocks as there are particles of magnetite that are magnetic within some rocks and when rocks were formed, these magnetite particles aligned themselves with Earth's magnetic field. As the rocks cooled, the direction of the particles' magnetic polarity was fixed.

Also, the mantle occupies about 83% of the Earth`s volume and is the thickest of earth`s layers comprising large, dark igneous rocks with iron and magnesium as it extends down to about 2900km from crust to earth`s core. The mantle has a lower and upper solid layers; and the asthenosphere. The latter is characterized by a plastic behavior and flows slowly while it generates magma from its partial melting. The upper mantle and the crust make up the lithosphere, which is broken up into plates that move over the asthenosphere. The interaction of these plates is responsible for earthquakes, volcanic eruptions and the formation of mountain ranges and ocean basins. The crust is the Earth`s outermost layer that is made up of chemical elements such as silicon and aluminium. The continental crust vary in thickness between 20 km and 90 km, and an oceanic crust varies in thickness between 5 km and 10 km. The interior temperature of the Earth increases with depth through the transfer of heat to the surface by convention and conduction through liquid and solid layers, respectively.

The movement of the Earth's crust creates pressures in certain regions of rocks, bending rocks and creating either upward or downward folds. At some times, rocks can break and change position. Continued pressure causes movement along some cracks. Such cracks, along which the rocks move, are called faults, accounting for earthquake occurrence where stresses occur between plate boundaries. An earthquake is the vibration of Earth caused by the sudden release of energy, usually located at a fault that involves the movement of blocks of rock along fractures. Following an earthquake there are usually adjustments along a fault line, which generate a series of smaller earthquakes known as aftershocks that continue for a few days after the occurrence of the initial earthquake. More than 900 000 earthquakes are recorded around the world each year. Tsunamis are caused mainly by earthquakes, but also by underwater landslides and volcanic eruptions. They are seismic ocean waves that travel at hundreds of kilometres per hour over very long distances. These seismic waves have enormous energies and in deep water, the crest of the waves is low at sea and move speedily. As these seismic waves reach shallower water, their speed slows and their energy is converted into an increase in the height of the wave. These waves, on reaching shore, may be 38 m or more in height and can therefore be very destructive. New rocks formed from Earth`s crust at one location also results in deterioration at other locations. Different types of rocks exist in the Earth`s crust but are all classified into three categories of igneous, metamorphic or sedimentary based on the mode of formation. Despite the aberrations in the Earth movement, one of the beneficial effect include the provision of building materials, fossil fuels, precious metals and minerals from mining of rocks and minerals. The changes that occur in the physical breakdown and chemical alteration of rocks on Earth`s surface are caused by weather through air and water (Beazley, 1980).

# Chapter Three

# Living Organisms Predisposed to Changes in Global Climate

## 3.0. Living Organisms Predisposed to Changes in Global Climate

The effects of changes in the Earth`s surface temperature that includes a change in global climate have been obvious over the past decades in natural ecosystems with a long-felt aftermath on human life, culture, economy, population species community and ecological regions (Lepetz *et al.*, 2009). The predicted pattern of global climate change is a major concern in many areas of socio- economic activities, such as agriculture and forestry, and is a major threat for biodiversity and ecosystem function (Lepetz *et al.*, 2009). According to IPCC (2007), climate change results from emission of greenhouse gases such as carbon dioxide, methane and nitrogen in the past century that will cause atmospheric warming. It now becomes important to adapt to future changes that might occur in climate.

It is evident that most carbon dioxide released from agriculture resulted from microbial decay, organic matter and burning of plant residues. Ruminants emit methane gas produced during fermentation of organic materials, among others. The microbial transformation of nitrogen in soils and manures generated nitrous oxide under wet conditions (Smith *et al.*, 2007).

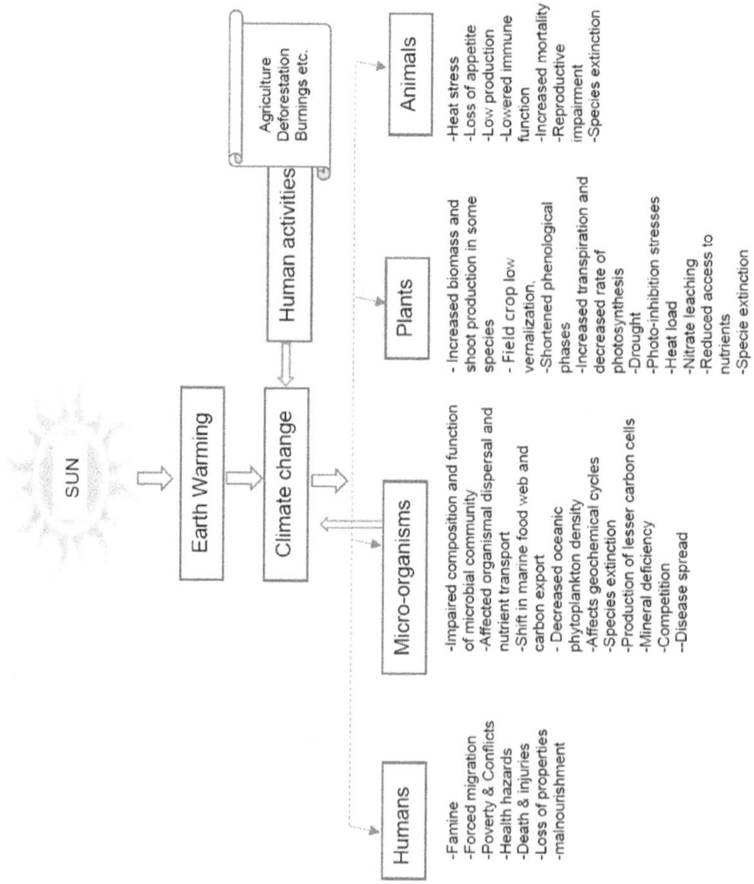


**Figure 2: Effects of climate change on living organisms**

## 3.1. Plants

Although plants are grouped according to their photosynthetic metabolic pathways, yet they are all affected by changes in climate. Around 95 % of the world plant biomass are grouped in C3, C4 and Crassulacean acid metabolism (CAM) plant species which are based on enzymes involved in photosynthetic fixation of carbon dioxide ($CO_2$), namely Rubisco, carboxylase and to some extent carbonic anhydrase, which are significantly different in their response to $CO_2$ enrichment. The $CO_2$ and other minerals can activate Rubisco by binding at a non-catalytic site on the enzyme protein. The process of photorespiration rate is high in C3 plants and the relative proportion of $CO_2$ and $O_2$ inside the leaf determine the rate of photorespiration. Under high carbon dioxide condition, both C3 and C4 species show increase in the total plant biomass of 44% and 33% respectively. While C3 species increased in tiller formation, C4 species increased in leaf area. Net assimilation of $CO_2$ increased in both C3 and C4 plant species, while stomatal conductance of $CO_2$ and water vapor decreased for C3 and C4 species by 24% and 29%, respectively (Wand *et al.*, 1999).

Increased biomass and shoot production were observed in both C3 and C4 plants under elevated $CO_2$, and also varies with precipitation and temperature (Chen *et al.*, 1996). The growth of plants in high atmospheric $CO_2$ condition has decreased the vulnerability of such plants to drought and also enhances plant resistance to heat and frost stresses, by likely reflecting greater concentrations of membrane stabilizing sugars in the tissues and inducing greater nutrient deficiency, and as observed in several studies it leads to accumulation of secondary carbon-rich chemicals such as tannins (Niinemets, 2010).

Field crops also produce larger number of mesophyll cell, chloroplasts, good and extended roots under elevated $CO_2$ conditions (Qaderi and Reid, 2009). According to IPCC (2007), crop productivity increased under increased average temperature range of 1 to 3°C, but decreases above this range due to low vernalization, shortened phenological phases, increased transpiration and decreased rate of photosynthesis (Qaderi and Reid, 2009). However, Chen *et al.* (1996) noted that the differences in responses in different ecosystems to elevated $CO_2$ might be due to difference in water, soil nutrient availability and variations in ambient temperature, while Easterling and Apps (2005) affirmed using spring wheat that an increase in yield was observed in supply of water, while cotton crop exposed to carbon dioxide rich air had yield increase of about 48% and 37% of biomass.

Bolin *et al.* (1989) noted that the forest ecosystems and their interaction with climate variability is complex as a result of physical, biological, and chemical processes occurring in nature. An increase in the ambient $CO_2$ concentration could reduce the opening of stomata required to allow a given amount of $CO_2$ to enter in the forest trees that could reduce transpiration and thereby increasing the efficiency of water use and productivity. Although, Saxe *et al.* (2001) affirmed that trees are capable of adjusting to a warmer climate, yet they are specie dependent and is also a function of the photo-inhibitory and respiratory processes. As forest trees are characterized by the C3 photosynthetic path way, their productivity and demand for nutrient is highly affected by atmospheric $CO_2$ concentration and temperature. Lukac *et al.* (2010) observed that the total productivity expected from trees growing under elevated $CO_2$ is larger than that expected in crops. However, Tylianakis *et al.* (2008) opined that the estimation of production increase in forest trees which is assumed to be higher than in crops is dependent on the attainment of equilibrium between nutrient absorption and use efficiency. However, the long-term response of forest to rising level of carbon dioxide is still uncertain, though a positive response have been observed from several researches as affirmed by Norby *et al.* (2005), while increased temperatures could amount to heat load and water logging stress in bogs, thereby leading to more severe heat, drought and photo-inhibition stresses (Niinemets, 2010). Temperature affects photosynthesis by altering enzyme activity, electron transport and leaf temperature that influences stomatal conductance. An indirect temperature response occur when evaporation increases, with stomata closing to reduce water loss through transpiration leading to reduction in $CO_2$ assimilation rate due to less $CO_2$ supply to chloroplast (Lloyd and Farquhar, 2008).

Porter and Semenov (2005) also noted that climatic variability affects crop development and yield through linear and non-linear response to weather variables and exceeding of well-defined crop thresholds, particularly, temperature. In the processes of plant growth, leaf litter fall to soil ground and under different temperature conditions, various processes occur in and out of the soil surface, alongside other processes directly affected by increased temperature, as well as decomposition, weathering and mass flow diffusion, may be faster in the soil under optimum soil moisture condition. At low temperatures, the reaction processes become slower, as temperature can indirectly affect plant morphology, growth, roots turn over, if it is lower than the optimum level for plants. In addition, soil moisture, nutrient availability among others play important role in plant growth and development, with enhanced $CO_2$ and anthropogenic nitrogen improving plant growth rate at a short term phase, while changing plant chemical processes and affecting biotic interactions.

Agricultural practices have contributed to greenhouse gas emission and have contributed to changes in climate, especially under intensive system of farming due to the utilization of high amount of chemical inputs and changing human activities that varies according to different regions of the world. Vulnerability to climate change depends not only on physical and biological responses but also on socio economic characteristics that are specific to different regions. Low- income population especially those who cultivate crops under rain fed and non- irrigated agriculture systems in dry lands, arid and semi-arid areas highly affected by severe hard ship due to climate change (Grasty, 1999). Due to impacts of climate change, agricultural productivity have been directly affected in developing nations of the world (Alexandrov and Hoogenboom, 2000). Variations in weather patterns also determine the trend of food production. Events such as drought, precipitation, flood, extreme deforestation among others have also contributed to the reduction in agricultural productivity (Cerri *et al.*, 2007).

Soil warming can affect the availability of nutrients, increase soil nitrogen mineralization and nitrate leaching, while increasing the rate of organic matter decomposition. Several studies have shown that soil warming can affect availability of nutrient, increase soil nitrogen (N) mineralization, nitrate leaching, and organic matter decomposition, with a slight temperature increase producing significant enhancement of activities. An increase in nitrogen mineralization in the soil under favourable moisture conditions and substrate availability, is predictable mainly in those ecosystems where temperature is a limiting factor, which leads to increased net primary production. An increasing temperature will also enhance the release of nutrients locked up in organic soil fraction, while decreasing soil moisture may limit this process. A higher rate of weathering of nutrient rich rocks generally leads to higher base saturation of the soil and maintain higher soil pH, both characteristics favourable to plant growth (Lukac *et al.*, 2010). Temporal variability in environment is commonly believed to increase the probability of population extinction, most especially when environmental variability increases as a result of climate change. Thuiller *et al.* (2008) affirmed that climate change have already driven the extinction of rear edge plant populations leading to the distribution of a trailing edge.

According to Hardy (2003), tree cover provides a habitat for numerous herbaceous plants, fungi, lichens, small and large animals, therefore the destruction of such would undoubtedly affect all species that make up a complex forest ecosystem. Even though plants have their own natural mechanisms to tolerate some level of adverse environmental conditions, physiological responses of forest trees under climate change condition are highly determined by the limiting factors of a particular site of forest growth. For example, increasing temperature may also increase vapor pressure deficit (VPD) of the air, and thereby increase transpiration rates that may result in adverse effects, especially on dryer sites (Boyd, 2010). The growth in human population and industrialization have also led to a continuous resource competition over land use systems, and deforestation which have impacted negatively on biodiversity of different habitats, forest cover growth, and ecosystem services.

According to FAO (2007) and Minura (2010), climate change impacts are classified into two broad categories: (1) Biophysical impacts that indicates the physical impacts caused by climate change directly on physical environment such as drought, deforestation and flooding, among others and its aftermath is reflected in the quality and quantity of crops, pasture, forest, natural resources, increased weed, insect pest challenge, livestock and shift in spatial and temporal distribution of impacts; and (2) Socio economic impacts as evidenced in reduced yield and production, reduced marginal gross domestic product (GDP) from agriculture sector, fluctuation of world market price, change in geographical distribution of trade regimes, due to shortage of food in quality and quantity the number of people in hunger and risk increased, and uncontrolled migration.

According to Morris (2010), climate change affects species indirectly by reducing the amount and availability of habitats and by reducing species to species interaction frequency. As a result, the loss of one species can result in decrease, increase or extinction of other apparently unconnected species. However, indiscriminate human activities have resulted in extinction at higher level than expected from random species losses. When species are lost from an ecosystem, it is not the only species that is lost, but the interaction and the general ecological functions, which we expect from their interactions. The concentration of greenhouse gases (GHGs) in the atmosphere depends on the rate of chemical reactions in the atmosphere. The global warming potential of methane ($CH_4$) is 72 times higher than $CO_2$ over a period of 20 years and 25 times higher over a period of 100 years. However, since the current concentration of $CH_4$ in the atmosphere is much lower, compared to the level of $CO_2$, total warming effect of $CO_2$ is higher. Results from the anthropogenic GHGs emissions in 2004, revealed at global scale level, the amount of $CO_2$ emitted due to use of fossil fuel was 56.6%, and about 17% from biomass burning and deforestation, while other sources contributed about 2.8%. However, from total anthropogenic GHGs emission, $CO_2$ represented a sum of about 77%, and methane represented about 14 %, being produced from livestock, wet rice paddy field, different fermentation processes, energy and waste. Also, about 8% of nitrous oxide ($N_2O$) was produced from human activities in areas of agriculture and hydrocarbon fuel combustion, with hydroflurocarbon contributed about 1.1% that are produced from industrial activities, use of old air conditioners and refrigerators.

## 3.2. Micro-organisms

Microorganisms that are phototrophic in nature make use of the sun's energy in the top 200 m of the water column, whereas marine life in deeper zones uses organic and inorganic chemicals for energy (Jørgensen and Boetius, 2007). Marine community is influenced by sunlight and the availability of different forms of energy, and covers about 70% of Earth`s surface. The sun`s rising impact of temperature has affected biological processes but has also reduced water density leading to stratification and circulation, which affected organismal dispersal and nutrient transport. Also, precipitation, salinity and winds also affect stratification, mixing and circulation. The air and river bodies impact nutrients and significantly influence the composition and function of microbial communities, aside climate change. Flemming and Wuertz (2019) affirmed that the overall importance of microorganisms to oceanic ecosystems worth appreciating due to their number, function and biomass in the water column and sub-surfaces which is estimated to about 90% of the entire marine biomass. They also function within the ecosystem by fixing carbon and nitrogen, and re-mineralizing organic matter, thus forming the basis of oceanic food webs and global carbon cycle (Azam and Malfatti, 2007). The sinking, deposition and burial of fixed carbon in particulate organic matter to marine sediments is a key, long- term mechanism for sequestering $CO_2$ from the atmosphere. Therefore, the balance between regeneration of $CO_2$ and nutrients via re-mineralization seabed burial determines the effect on climate change. Aside the increased temperature of oceans, acidity in oceans increases as part of climate change effect and is predicted to change to about 0.4 units by the end of the century (Caldeira and Wickett, 2003). With the unprecedented rate of pH change, there is a need to understand the response of marine organisms to climate change (Riebesell and Gattuso, 2015). The effect of climate change on elevated greenhouse gas concentrations have transferred significant impact on ocean temperature, acidification, stratification, mixing, thermo-haline circulation, nutrient supply, irradiation and extreme weather events that collectively affect marine microbiota in ways that have substantial environmental consequences, including major shifts in productivity, marine food webs, carbon export and burial in the seabed (Hurd *et al.*, 2018). Behrenfeld (2014) noted that marine phytoplankton perform about half of the global photosynthetic $CO_2$ fixation and half of the oxygen production. Marine phytoplankton are distributed over a larger surface area, and are exposed to less seasonal variation with faster turnover rates compared to trees. Therefore, phytoplanktons respond rapidly on a global scale to climate variations. These characteristics are important when one is evaluating the contributions of phytoplankton to carbon fixation and forecasting production change in response to perturbations, though complicated by phytoplankton bloom cycles that are affected by both bottom- up control and top-down control. The bottom-up control involves nutrient availability and vertical mixing, while top down control might involve grazing and virus action (Hutchins and Boyd, 2016). According to Behrenfeld (2014), solar radiation, temperature and freshwater inputs to surface waters strengthen ocean stratification and consequently reduce transport of nutrients from deep water to surface waters, which reduces primary productivity. Although, increased $CO_2$ levels is beneficial to increasing primary production of phytoplankton, only when nutrients are not limiting (Verspagen *et al.*, 2014).

Boyce *et al.* (2010) reported that overall global oceanic phytoplankton density has decreased in the past century, but these conclusions have been questioned because of the limited availability of long- term variability in data on phytoplankton production and contrasting reports indicating an increase in oceanic phytoplankton production (Antoine *et al.*, 2005). As explained by Behrenfeld *et al.* (2017), the global sea ice is declining, leading to higher light penetration and potentially more primary production. However, there are conflicting predictions for the effects of variable mixing patterns and changes in nutrient supply and for productivity trends in polar zones, necessitating the appraisal and documentation of long- term data on phytoplankton production and microbial community composition. Long- term data are needed to reliably predict how microbial functions and feedback mechanisms will respond to climate change, yet only very few such datasets exist.

Boyd *et al.* (2019) noted that diatoms account for about 40% of particulate carbon export to depth and have relatively high sinking speeds compared with other phytoplankton groups. Physically driven seasonal enrichments in surface nutrients favour diatom blooms. However, anthropogenic climate change will directly affect these seasonal cycles, changing the timing of blooms and diminishing their biomass, which will reduce primary production and $CO_2$ uptake (Behrenfeld *et al.*, 2013). Through remote sensing data, a global decline of diatoms was documented between 1998 and 2012, particularly in the North Pacific, which was associated with shallowing of the surface mixed layer and lower nutrient concentrations (Rousseaux and Gregg, 2015). Chemolitho-autotrophic archaea and bacteria fix $CO_2$ under dark conditions in deep ocean waters and at the surface during polar winter and also contribute substantially to surface ocean respiration and cycling of many elements. Seafloor methanogens and methanotrophs are important producers and consumers of $CH_4$, but their influence on the atmospheric flux of greenhouse gas is uncertain (Boetius and Wenzhöfer, 2013). Marine viruses, bacteriovorous bacteria and eukaryotic grazers are also important components of microbial food webs; as they influence carbon sequestering and deposit into deep oceans (Guidi *et al.*, 2016). Bertagnolli and Stewart (2018) reported that climate change also affect global biological, geochemical cycles by influencing the interaction of predators and preys, including viruses. Oxygen minimum zones (OMZs) are global sinks for reactive nitrogen and microbial production of $N_2$ and $N_2O$ and these have expanded in the past 50 years as a result of ocean warming, which reduces oxygen solubility and accounts for about 25–50% of nitrogen loss from the ocean to the atmosphere. They are also the largest pelagic methane reservoirs in the ocean and contribute substantially to open ocean methane cycling. The observed and predicted future expansion of OMZs may therefore considerably affect ocean nutrient, greenhouse gas prediction, and the distributions of oxygen- dependent organisms. Benthic microorganisms show biogeographic patterns and respond to variations in the quantity and quality of the particulate matter sinking to the seafloor. As a result, climate change is expected to particularly affect the functional processes of ammonia oxidation that deep- sea benthic archaea perform and associated biogeochemical cycles (Danovaro *et al.*, 2016).

Aerosols affect cloud formation, thereby influencing sunlight irradiation and precipitation, but the extent and manner to which they influence climate changes remain uncertain. Marine aerosols consist of a complex mixture of sea salt, non- sea-salt sulfate and organic molecules and can function as nuclei for cloud condensation, influencing the radiation balance and climate. Phytoplanktons emit dimethylsulfide and promotes cloud condensation through the activities of its derivate "sulfate" (Gantt and Meskhidze, 2013). However, Sanchez *et al.* (2018) opined that the knowledge of the activities of marine phytoplankton and their contribution to aerosols will allow better predictions of how changing ocean conditions will affect clouds and feedback on climate. Based on the affirmation of Flemming and Wuertz (2019), the atmosphere itself contains about 1022 microbial cells, and it will be valuable to be able to determine the ability of atmospheric microorganisms to grow and form aggregates as regards their influence on climate.

Climate change perturbs interactions between species and influences species to adapt, migrate or extinct from the community (Hutchins and Fu, 2017). Ocean warming, acidification, eutrophication and overuse result in the decline of coral reefs and may cause ecosystems shifts towards macro-algae and benthic cyanobacterial mats (Enochs *et al.*, 2015; De Bakker *et al.* 2017). The capacity for corals to adapt to climate change is strongly influenced by the responses of their associated microorganisms, including micro-algal symbionts and bacteria. The hundreds to thousands of microbial species that live on corals are crucial for host health, for example by recycling the waste products, by provisioning essential nutrients and vitamins and by assisting the immune system to fight pathogens. However, environmental perturbation or coral bleaching can change the coral microbiome rapidly. Such shifts undoubtedly influence the ecological functions and stability of the coral–microorganism system, potentially affecting the capacity and pace at which corals adapt to climate change, and the relationships between corals and other components of the reef ecosystem (Quigley *et al.*, 2018). Generally, microorganisms can disperse more easily than macroscopic organisms. Nevertheless, biogeographic distinctions occur for many microbial species, with dispersal and environmental factors strongly influencing community composition and function (Hanson, *et al.*, 2012).

However, Riebesell and Gattuso (2015) noted that microorganisms, such as bacteria, archaea and microalgae, with large population sizes and rapid asexual generation have high adaptive potential. Elevated $CO_2$ levels also disrupt phytoplankton species processes. However, increasing $CO_2$ levels provided a selective advantage to a toxic microalga "*Vicicitus globosus*," leading to disruption of organic matter transfer across trophic levels (Riebesell *et al.*, 2018). The marine cyanobacterial genus *Trichodesmium* responds to long- term duration of higher $CO_2$ levels with irreversible genetic changes that increase nitrogen fixation and growth, while photosynthetic green alga *Ostreococcus tauri*, responds to increased $CO_2$ levels by increasing growth, cell size and carbon to nitrogen ratios, with an observed changes in the population structure, ecotypes and niche occupation, thereby affecting the broader food webs and biogeochemical cycles (Schaum *et al.*, 2012; Hutchins *et al.*, 2015). Rather than producing larger cells, the calcifying phytoplankton species *Emiliania huxleyi* responds to the combined effects of elevated temperature, $CO_2$ levels and associated acidification by producing lesser carbon cells. However, for this species, overall production rates do not change as a result of evolutionary adaptation to higher $CO_2$ levels (Schlüter *et al.*, 2014).

Micro-organisms` responses to $CO_2$ changes differ between communities. A mesocosm study identified variable changes in the diversity of viruses that infect *E. huxleyi* when it is growing under elevated $CO_2$ levels, and noted the need to determine whether elevated $CO_2$ levels directly affected viruses, hosts or the interactions between them (Highfield *et al.*, 2017). These illustrations have pointed to the need to improve human understanding of evolutionary processes and incorporate that knowledge into predictions of the effects of climate change. Ocean acidification presents marine microorganisms with pH conditions well outside their recent historical range, thereby affecting their intracellular pH homeostasis. Species that are less adept at regulating internal pH will be more affected, and factors such as organism size, aggregation state, metabolic activity and growth rate influence the capacity for regulation (Flynn *et al.*, 2012). Lower pH causes bacteria and archaea to change gene expression in ways that support cell maintenance rather than growth. In mesocosms with low phytoplankton biomass, bacteria make use of more resources to pH homeostasis than bacteria in nutrient- enriched mesocosms with high phytoplankton biomass. Consequently, ocean acidification is predicted to alter cellular growth efficiency, carbon cycling and energy fluxes, consequently changing food web, with the biggest effects expected in the oligotrophic regions, which include most of the ocean (Bunse *et al.*, 2016). Experimental comparisons of *Synechococcus sp*. growth under both present and predicted future pH concentrations showed effects both on cyanobacteria and cyanophage viruses that infect those (Traving *et al.*, 2014).

Environmental temperature and latitude correlate with the diversity, distribution and/or temperature optimum of certain marine taxa, with models predicting that rising temperatures will cause a poleward shift of cold-adapted environments. However, optimum temperature of phytoplankton from polar and temperate waters was found to be substantially higher than environmental temperatures, and an eco-evolutionary model predicted that optimum temperature for tropical phytoplankton would be substantially higher than observed experimental values (Barton *et al.*, 2016). However, understanding how well microorganisms are adapted to environmental temperature and predicting how they will respond to warming requires assessments of more than temperature optimum, which is generally a poor indicator of physiological and ecological adaptation of microorganisms from cold environments (Cavicchioli, 2016).

Toseland *et al.* (2013) noted that environmental and physiological factors influence the responses and overall competitiveness of microorganisms in their natural habitat as illustrated in eukaryotic phytoplankton responding to elevated temperatures through increased protein synthesis while reducing cellular ribosome concentration. Eukaryotic phytoplankton has ribosomes that are rich in phosphate, while alterations caused by climate change to their nitrogen- to-phosphate ratio will affect resource allocation in the global ocean. According to Webster *et al.* (2018), careful attention must be paid to the quantification and interpretation of responses of environmental microorganisms to ecosystem changes and climate change-enhanced stresses as increased ocean temperatures, acidification and decreased nutrient supplies are projected to increase the extracellular release of dissolved organic matter from phytoplankton, with changes in the microbial loop possibly causing increased microbial production at the expense of higher trophic levels. Warming can also increase mineral deficiency such as iron limitation of nitrogen- fixing cyanobacteria, with profound implications for new nitrogen supplied to food webs of the future warming oceans (Thornton, 2014).

Terrestrial biomes are more abundant than marine biomes with estimate of about 100 folds and account for a large proportion of Earth's total biomass. Terrestrial plants perform roughly half of net global primary production (Bar- On *et al.*, 2018). Singh *et al.* (2010) reported that soils store about 2,000 billion tonnes of organic carbon, which is more than the combined pool of carbon in the atmosphere and vegetation, with its microorganisms regulating the amount of organic carbon stored in soil and those released back to the atmosphere; and indirectly influencing carbon storage in plants and soils through the provision of macronutrients that regulate the productivity of nitrogen and phosphorus. The mutual benefits understood in plants and mycorrhizal fungal have been established by Fellbaum *et al.* (2012), who noted that plants provide a substantial amount of carbon to their mycorrhizal fungal symbionts, and in many ecosystems, mycorrhizal fungi are responsible for substantial amounts of nitrogen and phosphorus acquisition by plants. Plants remove $CO_2$ from the atmosphere through photosynthesis and create organic matter that fuels terrestrial ecosystems. Conversely, autotrophic respiration by plants and heterotrophic respiration by microorganisms release $CO_2$ back into the atmosphere. Temperature influences the balance between these opposing processes and the capacity of the terrestrial biosphere to capture and store anthropogenic carbon. Therefore, warming is expected to accelerate carbon release into the atmosphere (Singh *et al.* (2010). While grasslands cover about 29% of the terrestrial surface (Hovenden *et al.*, 2019), non-forested, arid and semiarid regions make up about 47% and are important for the carbon budget while responding differently to anthropogenic climate change than forested regions that cover about 30% of the land surface, and contain about 45% of terrestrial carbon, and make up about 50% of terrestrial primary production, sequestering up to 25% of anthropogenic $CO_2$ (Bonan, 2008).

Lakes make up about 4% of the non-glaciated land area, and shallow lakes emit substantial amounts of $CH_4$. Decomposed plant material covers about 3% of the land surface and due to plant productivity exceeding decomposition, intact peatlands function as a global carbon sink and contain about 30% of global soil carbon (Bragazza *et al.*, 2013). In permafrost, carbon accumulation in organic matter far exceeds the respiratory losses, creating the largest terrestrial carbon sink as noted by Lupascu *et al.* (2014). Climate warming of 1.5 to 2°C is predicted to reduce permafrost by 28–53%, thereby making large carbon reservoirs available for microbial respiration and greenhouse gas emissions. Through the evaluations of the top 10 cm of soil and whole soil profiles to 100 cm deep, which contain older stocks of carbon; warming increases carbon loss to the atmosphere (Crowther *et al.*, 2016). van Gestel *et al.* (2018) explained that showcasing the differences in carbon loss between different soil sites will require a greater range of predictive variables, in addition to soil organic matter content, temperature, precipitation, pH and clay content. Nevertheless, predictions from global assessments of responses to warming indicated that terrestrial carbon loss under warming is causing a positive feedback that will accelerate the rate of climate change, particularly in cold and temperate soils, which store much of the global soil carbon.

Higher $CO_2$ levels in the atmosphere increase primary productivity and thus forest leaf and root litter which leads to higher carbon emissions due to microbial degradation (Sayer *et al.*, 2011). Bradford *et al.* (2016) stated that higher temperatures promote higher rates of terrestrial organic matter decomposition and the effect is not just a kinetic effect on microbial reaction rates but results from plant inputs stimulating microbial growth. Microbial community composition, density of dead wood, nitrogen availability and moisture also influence the rate of microbial activity that necessitates the Earth system model predictions of soil carbon losses through climate warming to incorporate local controls on ecosystem processes. In this regard, plant nutrient availability affects the net carbon balance in forests, with nutrient- poor forests releasing more carbon than nutrient- rich forests (Fernández- Martínez, 2014).

Microbial respiration may be lower in nutrient-rich forests as plants provide less carbon to rhizosphere microorganisms. Plants release about 50% of fixed carbon into the soil, which is available for microbial growth. In addition to microorganisms using exudates as an energy source, exudates can disrupt mineral–organic associations, liberating organic compounds from minerals that are used for microbial respiration, thereby increasing carbon release. The relevance of these plant–mineral interactions illustrates the importance of biotic–abiotic interactions, in addition to biotic interactions (plant– microorganism) when one is evaluating the influence of climate change (Keiluweit *et al.*, 2015). Tang and Riley (2015) documented that thermodynamic models that incorporate the interactions of microorganisms and secreted enzymes with organic matter and minerals have been used to predict soil carbon–climate feedbacks in response to increasing temperature. A study predicted more variable but weaker soil carbon–climate feedbacks from a thermodynamic model than from static models. The availability of soil organic matter for microbial degradation and long- term storage depends on factors including soil mineral characteristics, acidity, water availability, climate, and types of soil microorganisms. The nature of the organic matter, in particular substrate complexity, affects microbial decomposition. Furthermore, the microbial capacity to access organic matter differs between soil types. However, increasing atmospheric $CO_2$ levels are predicted to allow greater microbial decomposition and less soil retention of organic carbon (Sulman *et al.*, 2014).

Stevnbak *et al.* (2012) reported that elevated $CO_2$ concentrations enhanced competition for nitrogen between plants and microorganisms. Herbivores (invertebrates and mammals) affect the amount of organic matter that is returned to soil and in general, microbial biomass and activity. Climate change have resulted in overall alterations to global nitrogen and carbon cycles that reduce terrestrial carbon sequestration. Detritivores, such as earthworms influence greenhouse gas emissions by indirectly affecting plants. This process occur by increasing soil fertility and soil microorganisms, through the modification processes of burrowing, feeding and deposition of waste products. The anaerobic gut environment of earthworms harbours microorganisms that perform denitrification and produce $N_2O$. Earthworms enhance soil fertility, and their presence can result in net greenhouse gas emissions, although the combined effects of increased temperature and decreased rainfall on detritivore feeding and microbial respiration may reduce emissions (Lubbers *et al.*, 2013; Thakur *et al.*, 2018).

Melting and degradation of permafrost allow microbial decomposition of previously frozen carbon to release $CO_2$ and $CH_4$. However, coastal permafrost erosion will lead to the mobilization of large quantities of carbon to the ocean, with potentially large $CO_2$ emissions occurring through increased microbial remineralization causing a positive feedback loop that accelerates climate (Jansson and Tas, 2014). Melting of permafrost leads to increases in water- saturated soils and promotes anaerobic methane production by methanogens and $CO_2$ production by a range of microorganisms. Production is slow compared with metabolism in drained aerobic soils, which releases $CO_2$ rather than $CH_4$. However, a 7-year laboratory study of $CO_2$ and $CH_4$ production found that once methanogen communities became active in thawing permafrost, equal amounts of $CO_2$ and $CH_4$ were formed under anoxic conditions, and it was predicted that by the end of the century, carbon emissions from anoxic environments will drive climate (Knoblauch *et al.*, 2018).

In natural and artificial anaerobic environments, methanogens also produce methane in addition to the anthropogenic methane production associated with fossil fuels. The main sinks for methane are atmospheric oxidation and microbial oxidation occurring in soils, sediments and water (Gålfalk *et al.*, 2016). van Groenigen *et al.* (2013) affirmed that in recent times, atmospheric methane levels have risen sharply from increased emissions from methanogens, fossil fuel from industries and reduced atmospheric methane oxidation, thereby posing a major threat to controlling climate warming. In an illustration, rice feeds half of the global population and rice paddies contribute about 20% of agricultural methane emissions despite covering only about 10% of arable land. Anthropogenic climate change is predicted to double methane emissions from rice production by the end of the century. Although, Ripple *et al.* (2014) noted that ruminant animals are the largest single source of anthropogenic methane emissions, with about 48 times larger carbon footprint for ruminant meat production than plant- based, high protein foods. Also, the production of meat from non- ruminant animals such as pigs, poultry and fish produced about 3 to 10 times more $CH_4$ than high- protein plant foods. According to Itakura *et al.* (2013), agriculture is the largest emitter of the potent greenhouse gas $N_2O$, which is released by microbial oxidation and reduction of nitrogen. The enzyme $N_2O$ reductase in rhizobacteria and other soil microorganisms can also convert $N_2O$ to $N_2$. Climate change perturbs the rate at which microbial nitrogen transformations occur from decomposition to mineralization, to nitrification, to denitrification and fixation to release $N_2O$.

Bahram *et al.* (2018) reported that global assessment of top soil categorized food web into fungal- based soil food webs and bacterial based food web. The fungal –based food web are common in extensively managed farming systems like grasslands and are better able to adapt to drought than bacterial- based food webs, which are common in intensive systems. Although, both occupy specific niches, they respond differently to precipitation and soil pH, indicating that climate change would have differential impacts on their abundance, diversity and functions. Aridity is predicted to increase owing to climate change, and this reduces soil microbial diversity and the overall functional potential of microbial communities, and consequently limiting their capacity to support plant growth

Posch *et al.* (2012) noted that the combined effect of climate change and eutrophication caused by fertilizers can have an unpredictable effect on microbial competitiveness. For example, nutrient enrichment typically favours harmful algal blooms, but a different outcome was observed in the relatively deep Lake Zurich. Reducing phosphorus inputs from fertilizers reduced eukaryotic phytoplankton blooms but increased the nitrogen- to-phosphorus ratio and thus the non-nitrogen-fixing cyanobacterium, *Planktothrix rubescens* became dominant. The authors noted further that in the absence of effective predation, annual mixing has an important role in controlling cyanobacterial populations. However, warming increased thermal stratification and reduced mixing, thereby facilitating the persistence of the toxic cyanobacteria.

### 3.2.1. Disease spread in microbial community

Climate change affects the occurrence and spread of diseases in marine and terrestrial biota, depending on diverse socioeconomic, environmental and host–pathogen- specific factors (Altizer *et al.*, 2013). Johnson *et al.* (2015) established that the understanding of the spread of disease and designing effective control strategies require knowledge of the ecology of pathogens, their vectors, hosts, influence of dispersal and environmental factors, while affirming that there is a strong link between increasing sea surface temperatures and coral disease and, although the disease mechanisms are not absolutely clear for all the different syndromes, associations with microbial pathogens exist (Randall and van Woesik, 2015).

Ocean acidification may also directly cause tissue damage in organisms such as fish, leading to a weakened immune system that creates opportunities for bacterial infection (Frommel *et al.*, 2012). Sea star species decreased by 80 to 100% along a section of the North American west coast, with peak decline occurring during excessive sea surface temperature increase. As sea stars are important predators of sea urchins, loss of predation can cause a trophic cascade that affects kelp forests and associated marine biodiversity (Harvell *et al.*, 2019). Maynard *et al.* (2016) noted that with the effects that ocean warming has on pathogen impacts, temperature monitoring systems have been developed for a wide range of marine organisms, including corals, sponges, oysters, lobsters and other crustaceans, sea stars, fish and sea grasses. Forest die- off caused by drought and heat stress can be exacerbated by pathogens. For crops, a variety of interacting factors are important when one is considering response to pathogens, including $CO_2$ levels, climatic changes, plant health and species- specific plant– pathogen interactions. A broad range of microorganisms such as fungi, bacteria, viruses, viroids and oomycetes cause plant diseases and can, therefore, affect crop production, with consequence of food shortage/famines (Bebber *et al.*, 2013). Raffel *et al.* (2013) observed that climate change can enhance disease risk by altering host and parasite acclimation. For ectotherms such as amphibians, temperature can increase susceptibility to infection, possibly through perturbation of immune responses. An example is the frequent unpredictable environmental temperature fluctuations that increases the susceptibility of the Cuban tree frog to the pathogenic chytrid fungus *Batrachochytrium dendrobatidis*. The effect of increasing temperature on infection contrasts with decreased growth capacity of the fungus in pure culture, illustrating the importance of assessing host–pathogen responses when evaluating the relevance of climate change. MacFadden *et al.* (2018) noted that climate change is predicted to increase the rate of antibiotic resistance of some human pathogens such as *Escherichia coli*, *Klebsiella pneumoniae* and *Staphylococcus aureus* by 2 to 10%, depending on the type of antibiotics, if an increase of daily minimum temperature of 10 °C occur. The possible explanation for these occurrences could have been the underlying mechanism in which elevated temperatures facilitates horizontal gene transfer of mobile genetic elements of resistance, and increased pathogen growth rates, promoting environmental persistence, carriage and transmission. Population growth, which amplifies climate change, is also an important factor in contributing to the development of resistance. Patz *et al.* (2005) stated that vector- borne, foodborne, airborne, waterborne and other environmental pathogens may be particularly susceptible to the effects of climate change. While assessing vector- borne diseases, climate change will affect the distribution of vectors and the range over which diseases are transmitted, as well as the efficiency with which vectors transmit pathogens. Efficiency depends on the time between a vector feeding on an infected host and the vector becoming infectious itself. At warmer temperatures, this time can be reduced substantially, providing more opportunity for transmission within the vector's lifespan. Certain vector- borne diseases, such as bluetongue, an economically important viral disease of livestock, have already emerged in Europe in response to climate change, and more frequent outbreaks with larger impact are predicted to occur in the future (Jones *et al.*, 2019).

For some waterborne infections by pathogenic *Vibrio* spp., poleward spread correlates with increasing global temperature, sea surface temperatures and lower salinity of aquatic environments in coastal regions due to higher precipitation, leading to *Vibrio* spp. growth promotion (Baker- Austin *et al.*, 2013). The spatial distributions of malaria and dengue fever which are vector- borne diseases, are expected to shift in response to climate change because they are known to be highly sensitive to climate conditions (Pecl *et al.*, 2017). According to Semenza and Domanovic (2013), climate change can facilitate the spread of vector-borne pathogens by prolonging the transmission season, increasing pathogens` replication rate in the vector and increasing the number and geographic range of mosquitoes. This is especially the case for *Aedes aegypti*, the major vector of dengue, Zika, chikungunya and yellow fever viruses, which is currently limited to tropical and subtropical regions because it cannot survive cold winters. In combination with other mosquito- borne diseases such as West Nile fever and Japanese encephalitis and tick- borne diseases such as Lyme disease, millions of people are predicted to be newly at risk under climate change.

Adaptation of micro-organisms species to changes in their local environment has been less studied compared to animals (including humans) and plants. Although, the mechanisms and consequences of adaptation have been studied in natural and experimental microbial populations, viral, bacterial and fungal pathogens of plants and animals such as humans and livestock, adapt to abiotic and biotic factors such as temperature, pesticides, interactions between microorganisms and host resistance, in ways that affect ecosystem function, human health and food security (Giraud *et al.*, 2017). Croll and McDonald (2017) reported that the cyclic feedback between microbial response and human activity is well illustrated by the adaptation patterns of pathogenic agricultural fungi and because agricultural ecosystems have common global features, human travels, transport of plant materials that readily disperse crop pathogens, and agriculturally adapted pathogens have higher potential to cause epidemics and pose a greater threat to crop production than naturally occurring strains. The ability of fungal pathogens to expand their range and invade new habitats by evolving tolerant traits for higher temperatures aggravates the threat that they pose to both natural and agricultural ecosystems (Robin *et al.*, 2017).

## 3.3. Humans and ecosystem networks

An ecosystem is a dynamic complex system of plant, animal, microorganism communities and the non-living environment interacting with each other as a functional unit. Ecosystem services or networks are the benefits that people get from ecosystems, like food, forest products, water quality and quantity, soil conservation, biodiversity, recreation, and other cultural values. In order to get good ecosystem services for human wellbeing there should be a mechanism, which maintain the nutrient cycles and production, in a way to enhancing sustainability and conservation of natural resources. Any disruption or loss of natural ecosystems leads to breakdown of ecosystem functions and loss of ecosystem services. Climate change results in extreme events of drought and forest fire, endangering large proportion of species, with some being extinct. Also, ecosystem services includes water purification by wetlands, the protection of coastal areas from storm surges by mangroves and coral reefs, the regulation of pests and diseases, recycling of waste nutrients, and the removal of carbon from the atmosphere (Warren, 2011).

According to Bolin *et al.* (1989), a fundamental difference between global ecosystems of the past and the future is the dominating influence of human activity and intervention in natural environment. In addition, deforestation, agriculture and over grazing can hasten the processes of desertification especially in sub tropics and semiarid lands. The reoccurring human activities has resulted in climate change and shift in environmental balance, creating unprecedented extinction of plants and animals species, loss of biodiversity and specie threat. Although, human effects on microorganisms are less obvious and certainly less characterized, a major concern is that changes in microbial biodiversity and activities will affect the resilience of all other organisms and hence their ability to respond to climate change live in all environments on earth that are occupied by macroscopic organisms, and they are the sole life forms in other environments, such as the deep subsurface and extreme environments. Microorganisms date back to the origin of life on Earth at least 3.8 billion years ago, and they will likely exist well beyond any future extinction events. Although microorganisms are crucial in regulating climate change, they are rarely the focus of climate change studies and are not considered in policy development. Their immense diversity and varied responses to environmental change make determining their role in the ecosystem challenging.

Some studies particularly on agricultural crops and forest have shown that the enhanced atmospheric $CO_2$ directly increase productivity by stimulating net photosynthetic activity. Transpiration decreases through a partial stomatal closure resulting in increased water use efficiency of plants at least at a leaf scale. However, these responses are specie mediated as some species in terrestrial ecosystems may in the long-term indirectly react negatively, perhaps fatal, to increased $CO_2$ concentration. The indirect responses of ecosystems are due to the effect of elevated $CO_2$ concentration brought about by changes in temperature or radiation, humidity, precipitation or other climatic variables (Bolin *et al.*, 1989). As environmental exploitations by humans are increased, the global environmental change, including increased atmospheric $CO_2$ levels, associated climate change fragmentation and loss of natural habitats will also increase, resulting into the world`s rapid change in ecosystems.

Tylianakis *et al.* (2008) noted that the complex networks of biotic interactions such as predation, parasitism and pollination play an important role in the maintenance of biodiversity, mediation of ecosystem responses, and the stability of ecosystem services on which human well-being is dependent. The amount of atmospheric $CO_2$ concentration and ambient temperature determine the uptake of nutrients, minerals, water, plants` canopy exchange, and light absorption, needed for the formation and breakdown of carbohydrate through photosynthesis and respiration, respectively. The processes of transpiration are affected by the opening and closure of stomata and evaporation from plant surfaces and these are also determined by the amount of temperature and carbon dioxide. Soil biotic and abiotic processes are affected by climate change, sometimes resulting in a change in net primary production, species composition and aggressive competition for resources, consequently limiting the benefits of ecosystem services such as forest products, biodiversity, species composition, soil and water resources and recreation.

Higher evaporation and precipitation rates are also consequences of global warming and vulnerable areas includes regions humid tropics and higher latitudes with expectation of more precipitation while less precipitation is projected at mid-latitudes, semi-arid areas and dry tropics. The spatial extent of severe soil moisture deficits and frequency of short-term drought due to shortage and absence of rain is expected to double until late 21$^{st}$ century and long-term drought become three times more common especially in regions with less precipitation (IPCC, 2007).

The warming occurring on earth is not uniformly distributed as different continents show more rapid temperature increase when compared to oceans. Temperature change will have very different impacts on vegetation and ecosystem structure and productivity (Morison and Lawrol 1999). Global temperature has increased about 0.2°C per decade for the past 30 years and warming is larger in the western equatorial pacific than in the eastern equatorial pacific over the past century (IPCC, 2007). Regional warming has been predicted to increase with increasing latitude. However, warming in tropics is relatively small in its magnitude, but it is likely to have more consequences on ecosystems. These results imply that the greatest extinction risks from global warming may be in tropics, where biological diversity is also greatest (Fussel, 2009).

Temperature can affect photosynthesis through modulation of the rates of activities of photosynthetic enzymes and the electron transport chain and indirectly through leaf temperatures defining the magnitude of the leaf–to-air vapor pressure difference, which is a key factor influencing stomata conductance. Unlike the temperature sensitivity of processes like flowering and fruiting many other physiological processes have small genotypic variations, although some genetic adaptation have been observed on enzymes like Rubisco (Lloyd and Farquhar, 2008).

Microbial agents with food, water or vector-borne activities function in climate change by inducing infectious disease outbreak, with low infective doses and high virulence, and being environmentally stable and resistant to chlorine, iodine and salts. Such disease outbreak includes hepatitis A virus, gastroenteritis caused by the coccidian protozoans, astroviruses, caliciviruses, *Giardia lamblia*, the non-cholera marine vibrios such as *Vibrio vulnificus*, *Vibrio parahaemolyticus; Campylobacter jejuni*, Shigella species, and the Salmonella species involved in typhoidal and non-typhoidal illnesses. *Campylobacter jejuni* and many Salmonella species are natural commensals that live in the gastrointestinal (GI) tracts of birds, amphibians, and reptiles. Following natural disasters, man may be exposed to these bacteria following the consumption of contaminated drinking water or food exposed to animal or human carriers, or to contaminated fomites, such as cutting boards and carving knives. Humans can become carriers and fecal super shedders of *Salmonella typhi*. Although, these bacteria are very heat sensitive and can be destroyed by boiling drinkable water and thoroughly cooking meat, especially poultry. Following natural disasters of floods and tropical cyclones, reptiles are often displaced from their dens and can contaminate food preparation surfaces with non-typhoidal Salmonella species, particularly outdoor surfaces on which fresh fruits are later sliced for eating or squeezed or tapped for juices. On the other hand, the Vibrio bacteria, including Vibrio cholera, are natural sea water dwelling organisms that often contaminate human drinking water reservoirs following storm surges and can also contaminate open wounds and shell fish that are later consumed raw. Cholera-induced watery diarrhea can be treated with rehydration only, with antibiotics reserved for dysentery and prolonged or complicated infections. The non-cholera vibrios are much more ubiquitous and virulent pathogens than cholera vibrios, requiring antibiotic therapy in addition to rehydration, and can cause systemic septicemia and *Ecythema gangrenosa* with high case fatality rates, especially in those patients with pre-existing liver diseases, chronic alcoholism, hemochromatosis, or on iron supplementation. Examples of arthropod-borne viral disease outbreaks includes West Nile Virus in North America, St. Louis encephalitis, and LaCrosse encephalitis. Other outbreaks includes dengue, malaria, and yellow fever in the subtropical and tropical Americas; dengue and Japanese encephalitis in Asia, Indonesia, and Micronesia; and Ross River fever in Australia. As noted, malaria caused by chloroquine-resistant *P. falciparum* will occur at higher altitudes throughout Latin America, Africa, Asia, and the Middle East. Outbreaks of plague caused by *Yersinia pestis* and transmitted from rodents to man by fleas will continue to occur in regions where the organism is endemic in prairie dogs and other rodents, such as the US Four Corners (shared state boundaries of Arizona, Colorado, New Mexico, and Utah) and the southern Himalayas. Leptospirosis is caused by the spirochete, *Leptospira interrogans*, and is transmitted to man via rodent urine through contaminated surface waters occuring in coastal and other watersides. The large rodent populations in these cities will be frequently displaced by heavy rainfall, flash flooding, tidal and storm surges. Pulmonary hantavirus syndrome, another rodent zoonosis, transmitted by rodent fecal aerosols, will occur in dry areas following heavy rainfalls that force domestic deer mice from their underground burrows into vacant homes, garages, and crawl spaces. All the tick-transmitted diseases will shift their continental distributions northward by several latitudes and their altitude distributions higher by tens of meters.

Disasters that are water specific such as floods, rainstorm, tornadoes, tropical storms, cyclones, and earthquake-spawned tidal waves or tsunamis are commonly experienced by humans compared to geological disasters in Americas, Asia, and Africa and often results in more deaths by drowning than other disasters, and more serious injuries (Brewer *et al.*, 1994; Curriero *et al.*, 2001). The authors noted further that increased risks of endemic diseases, especially acute gastro-intestinal illnesses; and mass population movements are more common after water-centered disasters than after geological disasters, such as earthquakes and volcanic eruptions. However, Brewer *et al.* (1994) affirmed that storm surge is the greatest threat to human life during hurricanes and tsunamis, while storm tide is the greatest threat to property during hurricanes and tsunamis. A storm surge is a massive wall of seawater 50 to 100 miles wide that sweeps across the coastline near hurricane or tsunami landfall. A storm tide is the combination of the storm surge and the astronomical high tide. The storm tide sweeps along coastlines at hurricane or tsunami landfall for much longer distances (over 100 miles) and penetrates much deeper inland than the storm surge. Since hurricane Hugo in 1989, several surveillance investigations and needs assessments following tropical storms and hurricanes have been conducted by the US Centers for Disease Control and Prevention (CDC) and by local US and Latin American public health departments and ministries. Over the 2 weeks following hurricane Hugo in North Carolina, in 1989, 2,090 patients were treated in local hospital emergency departments (EDs) for injuries and illnesses related to the hurricane. Injuries were responsible for 88% of patient encounters, with half of the injuries due to lacerations and insect bites or stings. More than 50% of the lacerations were caused by chain saw injuries during the cleanup of fallen trees and limbs; 26% of the insect bites or stings were associated with systemic reactions, including anaphylaxis, wheezing, and hives. Following hurricane Andrew (Miami and Dade County, FL, USA, 1992), 25,000 homes were destroyed; 200,000 people were left homeless; 66% of the remaining households lost electricity; but most households had sewer service and non-potable running water and lots more. Six index conditions accounted for the most visits to civilian free-care sites: injuries amounted to 15.7%, miscellaneous infections amounted to 9.6%, nonspecific rash amounted to 5.4%, cough amounted to 4.7%, diarrhea amounted to 4.7%, and animal bites amounted to 1.2%. Five index conditions accounted for the most civilian visits to military free-care sites with 24.4% accounting for chronic disease conditions, 23.7% accounting for injuries, 12.4% accounting for dermatological conditions, 9.9% accounting for respiratory illnesses, and 5.3% accounting for diarrhea infection. During the 5 weeks after hurricane Andrew, the proportional morbidity from injuries decreased, proportional morbidity from respiratory illnesses increased, and proportional morbidity from diarrheal diseases remained stable, and there were no occurrence of infectious disease outbreak. The conclusion drawn was that injuries and exacerbations of chronic diseases were the most common causes of morbidity immediately after hurricanes and that infectious diseases were uncommon, and were not associated with communicable disease outbreaks (Lee *et al.*, 1992).

In a report of death associated with hurricanes Marilyn and Opal-United States, following hurricane Opal (Navarre, FL, USA, 1995), 1,135 patients were treated in the during the first 6 days after the hurricane and were compared to 996 patients treated by during a pre-hurricane period. For both periods, the proportion of visits were similar for lacerations, puncture wounds, musculoskeletal injuries, rashes, and gastro-intestinal and respiratory illnesses. During the post-hurricane period, however, the proportion of visits for insect bites and stings increased from 0.2% to 1.7%, in a mean age of 52.4 years for decedents, with most deaths occurring in males as a result of accidental injuries and three death records from acute exacerbations of chronic illnesses. A ranking of the causes of accidental deaths included falling and fallen trees, fewer cases of carbon monoxide poisoning from candle-started house fires, unventilated propane cooking tanks, unventilated gas powered electrical generators, drowning, a case of electrocution of a utility worker and massive chest trauma after a tractor overturned. The three deaths related to exacerbations of chronic diseases included two fatal myocardial infarctions in patients with coronary artery disease and one respiratory failure in a patient with chronic obstructive pulmonary disease. Early reports indicated that the most common causes of the 29 deaths during hurricane Isabel, a category 2 hurricane that made landfall on the North Carolina Outer Banks on September 18, 2003, pushing 6-foot storm surges and 12-foot storm tides north into Chesapeake Bay, included large cases of motor vehicle accidents, blunt injuries from falling tree limbs and utility poles, fewer cases of drowning and carbon monoxide poisoning from electrical power generator use in unventilated living spaces and a case of electrocution of electrical utility worker (Lee *et al.*, 1992).

Popoola *et al.* (2020a) noted that tropical regions across the globe are characterized by an annually prolonged hot weather conditions that showcase limiting production and economic efficiency in livestock industry, as domesticated animals exit the zone of maximum comfort, under heat stress, with death and reduced carcass yield accompanying the subsequent alteration in body chemistry and behavior. However, pen house orientation, cooling systems, genetic modification and different dietary manipulations have been employed in poultry industry, but many of such did not account for the body`s acid-base equilibrium and the potentials of aggregate levels of dietary electrolytes in replenishing lost electrolytes for enhanced nutrient utilization and carcass yield of poultry birds under severe heat stress conditions. Climate change has undoubtedly affected livestock industry, with a much more felt impact on poultry, especially meat and egg type fast-growing species; with an indirect impact on poultry products, shelf-life of feed ingredients, animal protein security and economic return. Direct impact have been showcased in the increasing trends of mortality, as an aftermath of thermal instability. The reduction in hen day egg production and qualities of egg external and internal features demands prompt attention. Pen house re-orientation with fortified air-conditioning systems still create different physiological and economic instability that is not commensurate to the 21st Century production agenda in different parts of the world. The most vulnerable region to climate change as reported by the Intergovernmental Panel on Climate Change (IPCC) have been small island states including the Caribbean.

Jones *et al.* (2015) affirmed through statistical means, a pronounced heat load predominance in the Caribbean for the periods of 1901–2012, 1951–2012 and for over half the region during 1979–2012. According to Stephenson *et al.* (2014), global warming existed between 1961–2010 and

1986–2010 using regional station data, with a greater warming impact with minimum temperature than maximum temperature, and an increase in warm day frequency, warm night frequency and decrease in cool days and night. Therefore, temperature will continue to increase until the end of the current century. Also, similar reports were observed with projections of temperature increase of about 3.5°C by the end of the century. However, Popoola *et al.* (2020b) affirmed that under severe heat stress conditions, with environmental temperatures exceeding 42 to 45°C, water intake of broiler birds at starter and finisher phases was not positively correlated to increase in body temperature. Therefore, a dietary balance of various electrolytes in diet must be made a common practice to correct blood acid-base and alleviate the adverse effects of heat stress (Popoola *et al.*, 2020c).

# Chapter Four

# Adaptation Measures for Changes in Global Climate

## 4.0. Adaptation Measures for Changes in Global Climate

The response of humans to climate change includes two basic measures: adaptation and mitigation. As defined by IPCC (2007), adaptation includes a set of continuous stream of activities or attitudes by individuals, society, and governments that informs decisions about all aspects of life for suitability of purpose by modifying processes. In ecology, adaptation indicates change in an organism′s physiology, behavior or other characteristics that increase the fitness to the environment, and to genetic changes. Cultural adaptation refers to adjustment by individuals and to the collective behavior of socio-economic systems and also includes changes in cognitions that are socially constructed and negotiated (Grothmann and Patt, 2003). Adaptation reflects existing social norms and processes, which can be motivated by many factors to moderate harm or exploit beneficial opportunities in respond to climate change. However, mitigation is an anthropogenic intervention to reduce sources of greenhouse gases (GHGs) or enhance sinks of carbon sequestration, for reduction of climate change impact and minimize adaptation challenges.

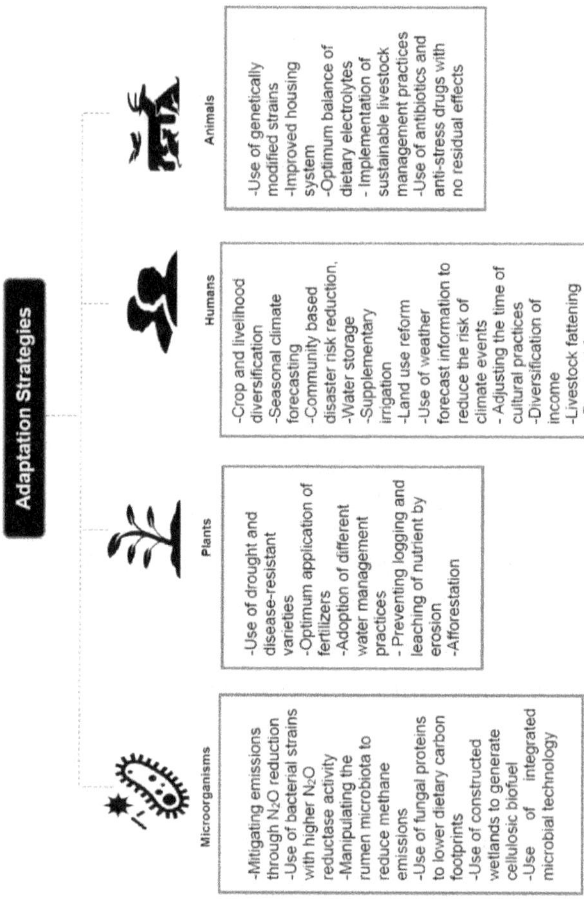


**Figure 3: Adaptation strategies to ameliorate the adverse effects of climate change on microorganisms, plants, humans and animals**

## 4.1. Microbial adaptation to changes in global climate

King *et al.* (2018) opined that an improved understanding of microbial interactions would help underpin the design of measures to mitigate the adverse effects of climate change. For example, understanding how mosquitoes respond to the bacterium *Wolbachia* (a common symbiont of arthropods) has resulted in a reduction of the transmission of Zika, dengue and chikungunya viruses through the introduction of *Wolbachia* into populations of *A. aegypti* mosquitoes and releasing them into the environment. Bakken and Frostegård (2017) also affirmed that the understanding of the ecology and physiology of microorganisms that reduce $N_2O$ to harmless $N_2$ provides options for mitigating emissions. The use of bacterial strains with higher $N_2O$ reductase activity has lowered $N_2O$ emissions from soybean, and both natural and genetically modified strains with higher $N_2O$ reductase activity provide avenues for mitigating $N_2O$ emissions as stated by Itakura *et al.* (2013). Manipulating the rumen microbiota and breeding programmes that target host genetic factors that change microbial community responses are possibilities for reducing methane emission from cattle, with the aim of producing cattle lines that sustain microbial communities and emits less methane without affecting the health and productivity of the animals (Roehe *et al.*, 2016).

Ritchie *et al.* (2018) observed that fungal proteins can replace meat protein, thus lowering dietary carbon footprints. Biochar is an example of an agricultural solution, produced from thermochemical conversion of biomass under oxygen limitation for broadly and indirectly mitigating microbial effects of climate change, and improves the stabilization and accumulation of organic matter in iron- rich soils (Weng *et al.*, 2017). The authors noted further that Biochar improves organic matter retention by reducing microbial mineralization and reducing the effect of root exudates on releasing organic material from minerals, thereby promoting growth of grasses and reducing the release of carbon. A potentially large- scale approach to mitigation is the use of constructed wetlands to generate cellulosic biofuel using waste nitrogen from wastewater treatment; if all waste in China were used, it could supply the equivalent of 7% of China's gasoline consumption (Liu *et al.*, 2012). Such major developments of constructed wetlands would require the characterization and optimization of their core microbial consortia to manage their emissions of greenhouse gases and optimize environmental benefits (Sánchez, 2017). Timmis *et al.* (2017) opined that microbial biotechnology can provide solutions for sustainable development, including in the provision of food and regulation of disease, and emissions and capturing of greenhouse gases of ecosystem services for humans, animals and plants. Also, microbial technologies provide practical solutions evolving round material, chemicals, energy and remediation for achieving many of the 17 United Nations Sustainable Development Goals, to address poverty, hunger, health, clean water, clean energy, economic growth, industry innovation, sustainable cities, responsible consumption, climate action, life below water, and life on land. A weighty support for such actions will undoubtedly be facilitated by improving public understanding of the key roles of microorganisms in global warming.

Glick *et al.* (2009) opined that adaptation to climate change entails ecological adaptation and its relation to the environment by ecosystem management and the change in social behavior to reduce the impacts of climate change. Thus, adaptation to climate change is the process through which people reduce the adverse effects of climate on their health and well-being, and take advantage of the opportunities that their climatic environment provides. Climate adaptation involves initiatives and measures to reduce the vulnerability of natural and human systems against actual or expected climate change effects (Glick *et al.*, 2009). Smit *et al.* (2000) stated another definition of adaptation, which involves adjustment to enhance the viability of social and economic activities and to reduce their vulnerability to climate, including its current variability and extreme events as well as long-term climate changes. According to Grothmann and Patt (2003), adaptation involves the action taken by people to avoid impacts or to benefit from opportunities associated with climate change. Natural resource conservationist propose the following definition for climate change adaptation for natural systems as a management strategy that involves identification, preparation for, and responses to expected climate change in order to promote ecological resilience, maintain ecological function, and provide necessary elements to support biodiversity and sustainable ecosystem services. In conclusion, climate change adaptation can be concluded to have the following characteristics of safe guarding, preparing for warming world, protecting wild life and natural resources from warm and coping mechanisms (Glick *et al*, 2009).

## 4.2. The need for adaptation

Adaptation is important in climate change issues that relates to the assessment of impacts and vulnerabilities; and the development and evaluation of response options (Grothmann and Patt, 2003). The danger and seriousness of climate change can be changed or reduced through the adoption of different kinds of practices, and making policies and strategic planning. During planning, the consideration of weather events and climate variables are important to prepare in advance and to decide the kind of measures, how and under what conditions adaptation practices can be performed (Smit *et al.*, 2000).

Depending on the degree of spontaneity, both adaptation and mitigation can effectively planned or performed autonomously. Planned adaptations largely relates with public agencies and autonomous adaptation with private sectors. Planned adaptation increases adaptive capacity by mobilizing institutions and policies to establish or strengthen conditions that are favourable to effective adaptation and investment in new technologies and infrastructures. Autonomous adaptation is the ongoing implementation of existing technologies and knowledge in response to changes in climate experiences (Rahman, 2008). Adaptation measures can take technological, economic, legal and institutional forms. For both assessment and implementation of adaptation, it is important to know the form of adaptation and the condition under which it might expect to occur (Smit *et al.*, 2000). It becomes both individual and community affair to deal with different extremes of weather events like flood, storms and drought. Autonomous adaptations are useful, especially for smallholder farmers, which includes using resistance varieties and different inputs, changing the rate of application of fertilizer to approve quality of agricultural products. Using different water management practices that includes improvement in soil water-holding capacity, physical soil and water conservation structures, and changing the amount and timing of irrigation, preventing logging and leaching of nutrient by erosion, some adverse effects of climate change can be ameliorated. The use of weather forecast information to reduce the risk of climate events, such as early warning about drought or outbreak of disease and pest in every season is also an alternative. Adjusting the time of cultural practices, diversification of income, livestock fattening, and other small agro-industries activities can help reduce the adverse effects of climate change. The use of technical options that includes many forms of land use, land use changes, new cultivation techniques have proven successful in adapting to climate change. Rahman (2008) also noted that planned approaches must include appropriate incentive structures, such as payment for environment services, in a way that can expand the options for indigenous people and poor communities to be beneficial in both mitigation and adaptation.

Adger *et al.* (2005) reported that adaptation includes different actions throughout society, individuals, groups and governments, which can be manifested and motivated by different factors, through actions of individuals, and organizations to meet targeted goals. Also, targeted goals can be set by government to protect the citizens; or by individual for their own benefit. Adaptation to climate variability aims to reduce vulnerability or increase resilience through the improvement in the ability to tackle and recover quickly from observed or expected climate change difficulties and weather events. Adaptation of physical, ecological and human systems include a change in social and environmental processes and practices, enabling a reduction in potential damage and finding new opportunities. Adaptation includes anticipatory and reactive actions to expected changes in temperature and climate variations and extremes. In practice, it should be on-going process that will reflect many stresses. For example, crop and livelihood diversification, seasonal climate forecasting, and many activities including community based disaster risk reduction, water storage, and supplementary irrigation, among others. Individuals can undertake some adaptation measures and others may be planned and implemented by government on behalf of the society. It is however expensive to implement adaptation measures (Adger *et al.*, 2007).

## 4.3. The extent of adaptation strategies

Climate change adaptation will be necessary in a variety of ecosystems, such as crops, livestock, grasslands, forestry, woodlands, and inland waters among others, and adaptation measures can differ according to several dimensions that includes spatial scale (local, regional and national), sectors (water resources, agriculture, tourism, and public health, etc); action type (physical, technological, investment, regulatory and market); stakeholders` practices (national, government, private sector, local communities, NGOs, and donors); climatic zone (arid land, dry land, arctic, flood plain or mountains etc); and income level (developed, middle and least developing countries). It also differs by the combination of these and other categories that includes current variability (learning from past adaptations to historical climate), observed medium and long term trends in climate and anticipatory planning in response to model-based scenarios of long term climate change. The research on vulnerability and adaptive capacity shows that some dimensions are generic, while others are specific to particular climate change impacts. Generic factors include education, income, health, and indicators specific to an impact such as drought or floods and may be relate to institution knowledge and/or technology. Technology play an important role in adapting to climate variability, for example seed improvement through plant breeding, cooling systems and other technologies and engineering solution are able to lead to improvement of the outcome and ability to cope with climate change conditions (Adger *et al.*, 2007).

The impact of climate change become serious on river and stream ecosystems in regions where water resources are already scarce and likely to become increasingly so due to high demand by growing human population. Furthermore, humans have made changes in some of rivers and streams, because of a number of factors including construction of dams, water diversions for agriculture and other structures, destruction of riparian vegetation and pollution of water, among others have significantly reduced the rivers and streams resistance capacity and their resilience to the impacts of climate change (Glick *et al.*, 2009).

## 4.4. Processes of adaptation

Stalker (2006) suggested that whatever the level of technologies that are available, its application is likely to be a more successive process which can be conducted in the following four main steps (i) adaptation measures needs to collect, interpret necessary information, and create awareness. (ii) Assessing and formulating different adaptation options that includes finding and assigning stakeholders to planning steps before implementation, and the design of appropriate and technically feasible policies that are consistent with the country′s development objectives.

The technology needs to be reasonably cost effective, environmentally sustainable, culturally compatible, and socially acceptable. (iii) Implementing actively supported by effective institutions, from national organizations up to individual communities. (iv) The technologies, which are already available needs to monitor, evaluate, follow up and make them more effective, and modified.

For the purpose of mitigation, transferring technologies (ideas or equipment) from developed countries to developing countries is necessary. However, technologies for adaptation differ from those for mitigation, because adaptation is not only concentrated in one sector, but also dispersed across all socio-economic sectors including agriculture, water, health, and infrastructure, so in a number of respects adaptation is divers and more complex than mitigation (Stalker, 2006). Rahman (2008) noted that adaptation measures can be site specific and target oriented, but if it is not properly planned and performed, it may cause a negative impact for the poorest people and those with insecure access to land. Similarly, some adaptation actions may cause a competition of resource, thereby affecting those who have no other alternative or access to irrigation schemes. Some adaptation measures may increase the demand of land that affects land less or small landowners, by making the price of the land higher. Such potential outcomes make adaptation complex and variable (Rahman, 2008). In addition, adaptation measures should be less capital intensive and more suitable to small scale interventions that needs to be more flexible and adaptable to local community and may therefore be more socially and legally accepted. However, as any type of technologies, there is always a risk that adaptation measures will be more accessible to rich communities. Therefore, Policy makers need to ensure that new forms of adaptation do not increase inequality, but rather contribute to a reduction in poverty (Stalker, 2006).

The benefits from researches widely shared and combined between local, modern techniques of genetic crop improvement and through improved management practice, using irrigation, and other inputs; have led to current yield improvement in some part of the world. Current technology should consider satisfying the interests of whole sectors including stakeholders, individual farmers, plant breeders, local communities, and other non-defined sectors. Although, Stalker (2006) noted that new technologies exist that are suitable for poor farmers in local conditions, it is somehow difficult to adapt, with small farm sizes and limited access to credit; procurement or investment in new technology may be difficult. According to Warren (2011), adaptation practice involves a cross-regional interaction between different regions, as the mechanism of climate change impact in one region can directly affect the other, through the loss of human or natural capital of one region affecting human or natural capital in another region, or indirectly through mitigation or adaptation measures practiced in one region can have a consequence for another. An example is agricultural yield reduction of a certain region causing an increase in the demand of imported food products from another area, with subsequent fluctuation in global price of food crop products. Changes in land use and land cover have large impact on climate change both locally and globally. For example, deforestation has led to surface albedo changed and the amount of carbon released from the soil, with increase in deforested vegetation. In addition, forests like Amazon recycle their water, thus, following forest loss, we may end up with drying. Climate change mitigation could involve significant reduction in deforestation, since it is regarded as the most cost effective way of reducing emission (Warren, 2011).

However, human adaptation to impacts of climate change might cause shift in agriculture from dry area to forest areas, and these can create additional impact on natural ecosystems. In approving food security, trees and shrubs in agro-forestry and farming systems can play a significant role in mitigating the impacts of extreme events (FAO, 2007). On the other hand, afforestation can also contribute to climate change mitigation through carbon sequestration, causing positive or negative implications for biodiversity and ecosystem services. Creating forests in areas where there is non-forest biodiversity using non-native tree species can have a negative impact on native or local biodiversity. Aside this, it might not succeed because of the soil and other climatic factors that may not be suitable for planting such non-native tree species. Mostly focusing on indigenous or local native trees in previously degraded or deforested areas will be beneficial for biodiversity and ecosystem services to enhance connectivity in forest ecosystems. However, it is important to notice that afforestation is a slowly process and it will therefore take a long time before a deforested area in terms of carbon storage can be compensated for by afforestation (Warren, 2011).

Existing programs and activities should consider the vulnerability to climate change and tailor these activities in adaptation and reduction in poverty. Agencies involved should focus on risk reduction works, through early warning signals. The risks though vary from place to place and between different groups, measures targeted at risks might be very specific to a particular situation. An example is seen in agricultural productivity reduction caused by climate change stress that could be targeted for solutions through adaptation measures, which will focus on changing cropping pattern techniques and using different modern technologies (Eriksen and O′Brien, 2007). Since different outcomes and adaptation exist, responses in a certain group may possibly affect the vulnerability context of some other groups somewhere else and adaptation responses might affect socio- environmental transformations. Sustainable adaptation is therefore a global environmental issue. In planning adaptation and decision-making processes, organizations have to focus on low risk knowledge, capacity building measures and specific actions to reduce vulnerability. However, the critical consideration is not only planning but also implementation of the adaptation processes (Eriksen, 2009).

## 4.5. Partners currently involved in adaptation and mitigation in vulnerable regions

Emergency departments relating to adaptation and mitigation of climate change, spread over a broad range of policy development and climatic variability strategies. At international level, different adaptation finance mechanisms established by the United Nations and the Kyoto Protocol includes Global Environment Trust Fund, Special Climate Change Fund, Least Developed Countries Fund, and Adaptation Fund; the first institutional adaptation plan has emerged rapidly from developed nations over the past six years, particularly at the state or province and local government or municipal scale (Preston *et al.*, 2010). However, partnership trends grow tremendously and has become not only organizational affair, but individual and household involvement.

Challenge of climate change requires coordinated action and united support of the international community as a whole. International organizations and other development agencies can bring their collective resources with the knowledge to support poor smallholder farmers. Such organization experience in working at field level with rural poor people on natural resource management and flood control, soil and water conservation, water-harvesting technologies and other projects in different part of the world, especially in arid and semi-arid regions, based on the principle of common, but differentiated responsibility will go a long way at local and national level to respond to climate change challenges through building capacities, by designing climate proof investments and mobilized resources. The Food and Agricultural Organization (FAO) focus on agricultural sectors, including forestry and can be relied upon as good and technically stable source of global information related to adaptations to climate change and extreme events. WFP, try to solve food insecurity by providing support during conflict and causes of climate related disasters, and private sectors are also the major source of funds for adaptation and mitigation (Rahman, 2008).

# Bibliography

Adger W. N., Arnell N. W., and Tompkins E. L. 2005. Successful adaptation to climate change across scales. *Global Environmental Change*, 15, 77-86.

Adger, W. N., Agrawala, S., Mirza, M.M.Q., Conde, C., O'Brien, K. , Pulhin, J., Pulwarty, R., Smitand, B. and Takahashi, K. 2007: Assessment of adaptation practices, options, constraints and capacity. Climate Change 2007: Impacts, Adaptation and Vulnerability. Contribution of Working Group II to the Fourth Assessment Report of the Intergovernmental Panel on Climate Change, M.L. Parry, O.F. Canziani, J.P. Palutikof, P.J. van der Linden and C.E. Hanson, Eds., Cambridge University Press, Cambridge, UK, 717-743.

Alexandrov, V. A. and Hoogenboom, G. 2000. Vulnerability and Adaptation Assessments of Agricultural Crops under Climate Change in the Southeastern USA; *Theor. Appl. Climatol.*, 67, 45-63.

Altizer, S., Ostfeld, R. S., Johnson, P. T., Kutz, S. and Harvell, C. D. 2013. Climate change and infectious diseases: from evidence to a predictive framework. *Science*, 341, 514–519.

Antoine, D., Morel, A., Gordon, H. R., Banzon, V. F. and Evans, R. H. 2005. Bridging ocean color observations of the 1980s and 2000s in search of long- term trends. *J. Geophys. Res. Oceans*, 110, C06009.

Azam, F. and Malfatti, F. 2007. Microbial structuring of marine ecosystems. *Nat. Rev. Microbiol.*, 5, 782–791.

Bahram, M. et al. 2018. Structure and function of the global topsoil microbiome. Nature 560, 233–237.

Baker- Austin, C. et al. 2013. Emerging Vibrio risk at high latitudes in response to ocean warming. *Nat. Clim. Change*, 3, 73–77.

Bakken, L. R. and Frostegård, Å. 2017. Sources and sinks for $N_2O$, can microbiologist help to mitigate $N_2O$ emissions? *Environ. Microbiol.*, 19, 4801–4805.

Bar- On, Y. M., Phillips, R. and Milo, R. 2018. The biomass distribution on Earth. Proc. Natl Acad. Sci. USA 115, 6506–6511.

Barton, A. D., Irwin, A. J., Finkel, Z. V. and Stock, C. A. 2016. Anthropogenic climate change drives shift and shuffle in North Atlantic phytoplankton communities. Proc. Natl Acad. Sci. USA, 113, 2964–2969.

Beazley, M. 1980. Anatomy of the earth, Mitchell Beazley, London.

Bebber, D. P., Ramotowski, M. A. T. and Gurr, S. J. 2013. Crop pests and pathogens move polewards in a warming world. *Nat. Clim. Change*, 3, 985–988.

Behrenfeld, M. J. 2014. Climate- mediated dance of the plankton. *Nat. Clim. Change*, 4, 880–887.

Behrenfeld, M. J. et al. 2017. Annual boom- bust cycles of polar phytoplankton biomass revealed by space- based lidar. *Nat. Geosci.*, 10, 118–122.

Behrenfeld, M. J., Doney, S. C., Lima, I., Boss, E. S. and Siegel, D. A. 2013. Annual cycles of ecological disturbance and recovery underlying the subarctic Atlantic spring plankton bloom. *Global Biogeochem. Cycles*, 27, 526–540.

Bertagnolli, A. D. and Stewart, F. J. 2018. Microbial niches in marine oxygen minimum zones. *Nat. Rev. Microbiol.*, 16, 723–729.

Boetius, A. and Wenzhöfer, F. 2013. Seafloor oxygen consumption fuelled by methane from cold seeps. *Nat. Geosci.*, 6, 725–734.

Bolin B., Döös B. R., Jager J., and Warrick R. A.,eds. 1989. The Green House Effect climate change and Ecosystems: SCOPE 29 International Council of Scientific Union. Great Britain.

Bonan, G. B. 2008. Forests and climate change: forcings, feedbacks, and the climate benefits of forests. *Science*, 320, 1444–1449.

Boyce, D. G., Lewis, M. R. and Worm, B. 2010. Global phytoplankton decline over the past century. *Nature*, 466, 591–596.

Boyd, J. 2010. Ecosystem service and climate adaptation, Resource for the Future, Issue brief 10-16.

Boyd, P. W., Claustre, H., Levy, M., Siegel, D. A. and Weber, T. 2019. Multi-faceted particle pumps drive carbon sequestration in the ocean. Nature 568, 327–335.

Bradford, M. A. et al. 2016. Managing uncertainty in soil carbon feedbacks to climate change. *Nat. Clim. Change*, 6, 751–758.

Bragazza, L., Parisod, J., Buttler, A. and Bardgett, R. D. 2013. Biogeochemical plant- soil microbe feedback in response to climate warming in peatlands. *Nat. Clim. Change*, 3, 273–277.

Brewer, R. D., Morris, P. D., and Cole, T. B. 1994. Hurricane-related emergency department visits in an inland area: an analysis of the public health impact of hurricane Hugo in North Carolina. *Ann Emerg Med.*, 23: 731 – 736.

Bunse, C. et al. 2016. Response of marine bacterioplankton pH homeostasis gene expression to elevated $CO_2$. *Nat. Clim. Change*, 5, 483–491.

Caldeira, K. and Wickett, M. E. 2003. Oceanography: anthropogenic carbon and ocean pH. *Nature*, 425, 365.

Cavicchioli, R. 2016. On the concept of a psychrophile. ISME J. 10, 793–795.

Cerri , C. E.P., Sparovek, G., Bernoux, M., Easterling, W.E., Melillo, J. M., and Cerri C. C. 2007. Tropical Agriculture and Global Warming: Impacts and Mitigation Options. *Sci. Agric. (Piracicaba, Braz.)*, v. 64, n. 1, pp., 83-89.

Chen, De. X., Hunt, H. W., and Morgan J. A. 1996. Responses of a C3 and C4 perennial grass to $CO_2$ enrichment and climate change: Comparison between model predictions and experimental data. *Ecological Modeling*, 87, 11-27.

Croll, D. and McDonald, B. A. 2017. The genetic basis of local adaptation for pathogenic fungi in agricultural ecosystems. *Mol. Ecol.*, 26, 2027–2040.

Crowther, T. W. et al. 2016. Quantifying global soil carbon losses in response to warming. Nature, 540, 104–108.

Curriero, F. C., Patz, J. A., Rose, J. B., and Lele, S. 2001. The association between extreme precipitation and waterborne disease outbreaks in the United States, 1948 – 1994. *Am. J. Pub. Health*, 91: 1194 – 1199.

Danovaro, R., Molari, M., Corinaldesi, C. and Dell'Anno, A. 2016. Macro-ecological drivers of archaea and bacteria in benthic deep- sea ecosystems. Sci. Adv., 2, e1500961.

De Bakker, D. M. et al. 2017. 40 years of benthic community change on the Caribbean reefs of Curaçao and Bonaire: the rise of slimy cyanobacterial mats. Coral Reefs, 36, 355–367.

Easterling W., and Apps M. 2005. Assessing the consequences of climate change for food and forest resources: a view from the IPCC: *Climate change*, 70, 165-189.

Enochs, I. C. et al. 2015. Shift from coral to macro-algae dominance on a volcanically acidified reef. *Nat. Clim. Change*, 5, 1083–1088.

Eriksen S. H. and O´Brien K. 2007. Vulnerability, poverty and the need for sustainable adaptation measures. *Climate Policy*, 7, 337-352.

Eriksen, S. 2009. Sustainable Adaptation Emphasizing Local and Global Equity and Environmental Integrity IHDP.

FAO of the United Nations, 2007. Adaptation to climate change in agriculture, forestry and fisheries: Perspective, framework and priorities; Viale delle Terme di caracalla-00100 Rome, Italy.

Fellbaum, C. R., Mensah, J. A., Pfeffer, P. E., Kiers, E. T. and Bücking, H. 2012. The role of carbon in fungal nutrient uptake and transport Implications for resource exchange in the arbuscular mycorrhizal symbiosis. *Plant Signal. Behav.*, 7, 1509–1512.

Fernández- Martínez, M. 2014. Nutrient availability as the key regulator of global forest carbon balance. *Nat. Clim. Change*, 4, 471–476.

Flemming, H. C. and Wuertz, S. 2019. Bacteria and archaea on Earth and their abundance in biofilms. *Nat. Rev. Microbiol.* 17, 247–260.

Flynn, K. J. et al. 2012. Changes in pH at the exterior surface of plankton with ocean acidification. *Nat. Clim. Change*, 2, 510–513.

Frommel, A. Y. et al. 2012. Severe tissue damage in Atlantic cod larvae under increasing ocean acidification. *Nat. Clim. Change*, 2, 42–46.

Fussel, M. H., 2009. An updated assessment of the risks from climate change based on research published since the IPCC Forth Assessment Report: Climate Change, 97: 467-482.

Gålfalk, M., Olofsson, G., Crill, P. and Bastviken, D. 2016. Making methane visible. *Nat. Clim. Change*, 6, 426–430.

Gantt, B. and Meskhidze, N. 2013. The physical and chemical characteristics of marine primary organic aerosol: a review. *Atmos. Chem. Phys.*, 13, 3979–3996.

Giraud, T., Koskella, B. and Laine, A. L. 2017. Introduction: microbial local adaptation: insights from natural populations, genomics and experimental evolution. *Mol. Ecol.*, 26, 1703–1710.

Glick, P., Staudt, A., and Stein B. 2009. A New Era for Conservation: Review of climate change adaptation literature: National Wildlife Federation.

Grasty, S. 1999. Agriculture and Climate Change: In TDRI Quarterly Review, Auger, P. and Suwanraks, R., eds.V.14.No.2 pp.12-16.

Grothmann, T., and Patt, A., 2003. Adaptive Capacity and Human Cognition: prepared for presentation at the Open meeting of the Global Environmental Change Research Community, Montreal, Canada, 16.

Guidi, L. et al. 2016. Plankton networks driving carbon export in the oligotrophic ocean. *Nature*, 532, 465–470.

Hanson, C. A., Fuhrman, J. A., Horner- Devine, M. C. and Martiny, J. B. H. 2012. Beyond biogeographic patterns: processes shaping the microbial landscape. *Nat. Rev. Microbiol.*, 10, 497–506.

Hardy, J. T. 2003. Climate change: cause, effects, and solutions, England: John Wiley & sons Ltd.

Harvell, C. D. et al. 2019. Disease epidemic and a marine heat wave are associated with the continental- scale collapse of a pivotal predator (*Pycnopodia helianthoides*). *Sci. Adv.*, 5, eaau7042.

Highfield, A., Joint, I., Gilbert, J. A., Crawfurd, K. J. and Schroeder, D. C. 2017. Change in *Emiliania huxleyi* virus assemblage diversity but not in host genetic composition during an ocean acidification mesocosm experiment. Viruses, 9, E41.

Hovenden, M. J. et al. 2019. Globally consistent influences of seasonal precipitation limit grassland biomass response to elevated $CO_2$. *Nat. Plants*, 5, 167–173.

Hurd, C. L., Lenton, A., Tilbrook, B. and Boyd, P. W. 2018. Current understanding and challenges for oceans in a higher- $CO_2$ world. *Nat. Clim. Change*, 8, 686–694.

Hutchins, D. A. and Boyd, P. W. 2016. Marine phytoplankton and the changing ocean iron cycle. *Nat. Clim. Change*, 6, 1072–1079.

Hutchins, D. A. and Fu, F. X. 2017. Microorganisms and ocean global change. *Nat. Microbiol.*, 2, 17508.

Hutchins, D. A. et al. 2015. Irreversibly increased nitrogen fixation in Trichodesmium experimentally adapted to elevated carbon dioxide. *Nat. Commun.*, 6, 8155.

IPCC, 2007: Summary for Policymakers. In: Climate Change 2007: Mitigation. Contribution of Working Group III to the Fourth Assessment Report of the Intergovernmental Panel on Climate Change [B. Metz, O.R. Davidson, P.R. Bosch, R. Dave, L.A. Meyer (eds)], Cambridge University Press, Cambridge, United Kingdom and New York, NY, USA.

Itakura, M. et al. 2013. Mitigation of nitrous oxide emissions from soils by *Bradyrhizobium japonicum* inoculation. *Nat. Clim. Change*, 3, 208–212.

Jansson, J. K. and Tas, N. 2014. The microbial ecology of permafrost. *Nat. Rev. Microbiol.*, 12, 414–425.

Johnson, P. T. J., de Roode, J. C. and Fenton, A. 2015. Why infectious disease research needs community ecology. *Science*, 349, 1259504.

Jones, A. E. et al. 2019. Bluetongue risk under future climates. *Nat. Clim. Change*, 9, 153–157.

Jones, P.D., Harpham, C., Harris, I., Goodess, C. M., Burton, A., Centella-Artola, A., Taylor, M. A., Benzanilla-Marlot, A., Campbell, J. D., Stephenson, T. S., Joslyn, O., Nicholls, K., and Baur, T. (2015). Long-term trends in precipitation and temperature across the Caribbean. *Int. J. Climatol.*, 36:3314–3333. https://doi.org/10.1002/joc.4557.

Jørgensen, B. B. and Boetius, A. 2007. Feast and famine —microbial life in the deep- sea bed. *Nat. Microbiol. Rev.*, 5, 770–781.

Keiluweit, M. et al. 2015. Mineral protection of soil carbon counteracted by root exudates. *Nat. Clim. Change*, 5, 588–595.

King, J. G., Souto- Maior, C., Sartori, L. M., Maciel-de- Freitas, R. and Gomes, M. G. M. 2018. Variation in Wolbachia effects on Aedes mosquitoes as a determinant of invasiveness and vectorial capacity. *Nat. Commun.*, 9, 1483.

Knoblauch, C., Beer, C., Liebner, S., Grigoriev, M. N. and Pfeiffer, E. M. 2018. Methane production as key to the greenhouse gas budget of thawing permafrost. *Nat. Clim. Change*, 8, 309–312.

Lee, L. E., Fonseca, V., Brett, K. M. et al. 1992. Active morbidity surveillance after hurricane Andrew-Florida, 1992. JAMA 1993; 270: 591 – 594.

Lepetz V., Massot, M., Schmeller, D.S., and Clobert, J., 2009. Biodiversity monitoring: some proposals to adequately study species' responses to climate change. Biodiversity and Conservation 18, pg 3185-3203.

Liu, D. et al. 2012. Constructed wetlands as biofuel production systems. *Nat. Clim. Change*, 2, 190–194.

Lloyd, J., and Farquhar, G. 2008. Effects of rising temperatures and [$CO_2$] on the physiology of tropical forest trees. *Phil. Trans. R. Soc. B*, 363, 1811-1817.

Lubbers, I. M. et al. 2013. Greenhouse- gas emissions from soils increased by earthworms. *Nat. Clim. Change*, 3, 187–194.

Lukac, M., Calfapietra, C., Lagomarsino, A. and Loreto, F. 2010. Global climate change and tree nutrition: effects of elevated $CO_2$ and temperature: Tree physiology [0829-318X] Lukac yr. 30, 1209 -1220.

Lupascu, M. et al. 2014. High Arctic wetting reduces permafrost carbon feedbacks to climate warming. Nat. Clim. Change 4, 51–55.

MacFadden, D. R., McGough, S. F., Fisman, D., Santillana, M. and Brownstein, J. S. 2018. Antibiotic resistance increases with local temperature. *Nat. Clim. Change*, 8, 510–514.

Maynard, J. et al. 2015. Projections of climate conditions that increase coral disease susceptibility and pathogen abundance and virulence. *Nat. Clim. Change*, 5, 688–694.

Mimura, N. 2010. Scope and roles of adaptation to climate change. In: A. Sumi, K. Fukushi, and A. Hiramatsu, eds., Adaptation and Mitigation Strategies for Climate Change. Springer Tokyo Berlin Heidelberg New York. Ch.9.

Monroe, J. S. and Wicander, R. 1994. The changing earth: exploring geology and evolution, West Publishing, London.

Morison, J. I. L and Lawlor D. W. 1999. Interaction between increase $CO_2$ concentration and temperature on plant growth. Plant Cell and Environment volume 22, 659-682.

Morris, R. J. 2010. Anthropogenic impacts on tropical forest biodiversity: a network structure and ecosystem functioning Perspective: Philosophical Transactions of the Royal Society. B 365, 3709-3718.

Niinemets, Ü. 2010. Responses of forest trees to single and multiple environmental stresses from seedlings to mature plants: past stress history, stress interactions, tolerance and acclimation. *Forest Ecology and Management*, 260, 1623-1639.

Norby R.J., Delucia E. H, Gielen B., Calfapietra C., Giardina C. P., King, J. S., Ledford J., McCarthy, H.R., Moore J.P., Ceulemans R., Angelis P. De, Finzil A.C., Karnosky D.F., Kubiske M.E., Lukac M., Pregitzer K.S., Scarascia-Mugnozza G. E., Schlesinger W. H., and Oren R. 2005. Forest response to elevated $CO_2$ is conserved across a broad range of productivity. PNAS 102, 18052-18056.

Norman, 2018. 10 Mind Blowing Facts About Africa.

Patz, J. A., Campbell- Lendrum, D., Holloway, T. and Foley, J. A. 2005. Impact of regional climate change on human health. Nature 438, 310–317.

Pecl, G. T. et al. 2017. Biodiversity redistribution under climate change: Impacts on ecosystems and human well- being. *Science*, 355, eaai9214.

Popoola, I.O., Popoola, O.R., Adeyemi, A.A., Ojeniyi, O.M., Olaleru, I.F., Oluwadele, F.J. and Akinwumi, E.O. 2020a. Overall Performance, Carcass Yield, Meat Safety Potentials and Economic Value of Heat-Stressed Broilers Fed Diets with Balanced Electrolytes. *Food and Nutrition Sciences*, 11, 615-628. https://doi.org/10.4236/fns.2020.117044.

Popoola, I.O., Popoola, O.R., Ojeniyi, M.O., Olajide, O.O. and Iyayi, E.A. (2020c). The Roles of Key Electrolytes in Balancing Blood Acid-Base and Nutrient in Broiler Chickens Reared under Tropical Conditions. *Natural Science*, 12, 4-11. https://doi.org/10.4236/ns.2020.121002

Popoola, I.O., Popoola, O.R., Olajide, O.O., Adeyemi, A.A. and Alegbejo, Q.T. (2020b). Reducing Sharp Fluctuations in Body Temperature and Optimizing Production Index of Broilers Using Dietary Electrolytes. *Open Journal of Animal Sciences*,10, 266-277. https://doi.org/10.4236/ojas.2020.102015

Porter, J. R. and Smenenov, M. A. 2005. Crop responses to climatic variation: *Phil. Trans. R.Soc.B* 360, 2021-2035.

Posch, T., Köster, O., Salcher, M. M. and Pernthaler, J. 2012. Harmful filamentous cyanobacteria favoured by reduced water turnover with lake warming. *Nat. Clim. Change*, 2, 809–813.

Preston, B. L., Westaway, R. M. and Yuen E. J. 2010. Climate adaptation planning in practice: an evaluation of adaptation plans from three developed nations: mitigation adapt strateg Glob change DOC 10.1007/s 11027-010-9270-x.

Qaderi, M. M. and Reid D. M. 2009. Crop Responses to Elevated Carbon dioxide and Temperature (chp1), In Singh S. N., (ed.), Climate Change and Crops, Environmental Science and Engineering, DOI 10.1007/978-3-540-88246-6 1, Springer-Verlag Berlin Heidelberg.

Quigley, K. M., Baker, A. C., Coffroth, M. A., Willis, B. L. and van Oppen, M. J. H. 2018. In Coral Bleaching: Patterns, Processes, Causes and Consequences Ch. 6 (eds van Oppen, M. J. H. & Lough, J. M.).

Raffel, T. R. et al. 2013. Disease and thermal acclimation in a more variable and unpredictable climate. *Nat. Clim. Change*, 3, 146–151.

Rahman, A., 2008. Climate Change and the future of smallholder agriculture; how can rural poor people be a part of the solution to climate change? Discussion paper prepared for the round table on climate change at the third- first session of IFAD`S Governing Council: IFAD Policy Reference Group on Climate Change, IFAD: Rome, Italy.

Randall, J. and van Woesik, R. 2015. Contemporary white-band disease in Caribbean corals driven by climate change. *Nat. Clim. Change*, 5, 375–379.

Riebesell, U. and Gattuso, J. P. 2015. Lessons learned from ocean acidification research. *Nat. Clim. Change*, 5, 12–14.

Riebesell, U. et al. 2018. Toxic algal bloom induced by ocean acidification disrupts the pelagic food web. *Nat. Clim. Change*, 8, 1082.

Ripple, W. J. et al. 2014. Ruminants, climate change and climate policy. *Nat. Clim. Change*, 4, 2–5.

Ritchie, H., Reay, D. S. and Higgins, P. 2018. Potential of meat substitutes for climate change mitigation and improved human health in high- income markets. *Front. Sustain. Food Syst.*, 2, 16.

Robin, C., Andanson, A., Saint- Jean, G., Fabreguettes, O. and Dutech, C. 2017. What was old is new again: thermal adaptation within clonal lineages during range expansion in a fungal pathogen. *Mol. Ecol.*, 26, 1952–1963.

Roehe, R. et al. 2016. Bovine host genetic variation influences rumen microbial methane production with best selection criterion for low methane emitting and efficiently feed converting hosts based on metagenomic gene abundance. *PLOS Genet.*, 12, e1005846.

Rousseaux, C. S. and Gregg, W. W. 2015. Recent decadal trends in global phytoplankton composition. *Global Biogeochem. Cycles*, 29, 1674–1688.

Sanchez, K. J. et al. 2018. Substantial seasonal contribution of observed biogenic sulfate particles to cloud condensation nuclei. Sci. Rep., 8, 3235.

Sánchez, O. 2017. Constructed wetlands revisited: microbial diversity in the –omics era. *Microb. Ecol.*, 73, 722–733.

Saxe, H., Cannell Melvin G. R., Johnsen O., Ryan, M. G., and Vourlitis G. 2001. Tree and forest functioning in response to global warming. *New Phytologist*, 149: 369-400.

Sayer, E. J., Heard, M. S., Grant, H. K., Marthews, T. R. and Tanner, E. V. J. 2011. Soil carbon release enhanced by increased tropical forest litter fall. *Nat. Clim. Change*, 1, 304–307 (2011).

Schaum, E., Rost, B., Millar, A. J. and Sinéad, C. 2012. Variation in plastic responses to ocean acidification in a globally distributed picoplankton species. *Nat. Clim. Change*, 3, 298–302.

Schlüter, L. et al. 2014. Adaptation of a globally important coccolithophore to ocean warming and acidification. *Nat. Clim. Change*, 4, 1024–1030.

Semenza, J. C. and Domanovic, D. 2013. Blood supply under threat. *Nat. Clim. Change*, 3, 432–435.

Singh, B. K., Bardgett, R. D., Smith, P. and Reay, D. S. 2010. Microorganisms and climate change: terrestrial feedbacks and mitigation options. *Nat. Rev. Microbiol.*, 8, 779–790.

Smit, B., Burton, I., Klein, Richard J. T., and Wandel, J., 2000. An anatomy of adaptation to climate change and variability climate change 45, 223-225.

Smith, P., D. Martino, Z. Cai, D. Gwary, H. Janzen, P. Kumar, B. McCarl, S. Ogle, F. O'Mara, C. Rice, B. Scholes, O.& Sirotenko, 2007: Agriculture. In Climate Change 2007: Mitigation. Contribution of Working Group III to the Fourth Assessment Report of the Intergovernmental Panel on Climate Change [B. Metz, O.R. Davidson, P.R. Bosch, R. Dave, L.A. Meyer (eds)], Cambridge University Press, Cambridge, United Kingdom and New York, NY, USA.

Stalker, P. ed., 2006. Technologies for Adaptation to Climate Change; (UNFCCC): Bonn, Germany.

Stephenson, T.S., Vincent, L.A., Allen, T., et al. (2014). Changes in extreme temperature and precipitation in the Caribbean region, 1961–2010. *Int. J. Climatol.*, 34:2957–2971. doi: https://doi.org/10.1002/joc.3889.

Stevnbak, K. et al. 2012. Interactions between above- and belowground organisms modified in climate change experiments. *Nat. Clim. Change*, 2, 805–808.

Sulman, B. N., Phillips, R. P., Oishi, A. C., Shevliakova, E. and Pacala, S. W. Microbe- driven turnover offsets mineral- mediated storage of soil carbon under elevated $CO_2$. *Nat. Clim. Change*, 4, 1099–1102.

Tang, J. and Riley, W. J. 2015. Weaker soil carbon–climate feedbacks resulting from microbial and abiotic interactions. *Nat. Clim. Change*, 5, 56–60.

Thakur, M. P. et al. 2018. Reduced feeding activity of soil detritivores under warmer and drier conditions. *Nat. Clim. Change*, 8, 75–78.

Thornton, D. C. O. 2014. Dissolved organic matter (DOM) release by phytoplankton in the contemporary and future ocean. *Eur. J. Phycol.*, 49, 20–46.

Thuiller W., Albert, C., Araujo, M. B., Berry, P. M., Cabeza, M., Guisan, A.,Hickler, T., Midgley, G.F., Paterson, J., Schurr, F. M, Sykes, M. T., and Zimmermann, N. E. 2008. Predicting global change impact on plant species distributions: Future Challenges, Perspectives in Plant Ecology, Evolution and Systematic, 9, 137- 152. Rubel Foundation, Published by Elsevier GmbH.

Timmis, K. et al. 2017. The contribution of microbial biotechnology to sustainable development goals. *Microb. Biotechnol.*, 10, 984–987.

Toseland, A. et al. 2013. The impact of temperature on marine phytoplankton resource allocation and metabolism. *Nat. Clim. Change*, 3, 979–984.

Traving, S. J., Clokie, M. R. and Middelboe, M. 2014. Increased acidification has a profound effect on the interactions between the cyanobacterium, *Synechococcus sp.* WH7803 and its viruses. *FEMS Microbiol. Ecol.*, 87, 133–141 (2014).

Tylianakis J. M., Didham R. K., Bascompte J., and Wardle D.A. 2008. Global change and spices interactions in terrestrial ecosystems: *Ecology Letters*, 11, 1351-1363.

van Gestel, N. et al. 2018. Predicting soil carbon loss with warming. *Nature*, 554, E4–E5.

van Groenigen, K. S., van Kessel, C., Hungate, B. A. 2013. Increased greenhouse-gas intensity of rice production under future atmospheric conditions. *Nat. Clim. Change*, 3, 288–291.

Verspagen, J. M., Van de Waal, D. B., Finke, J. F., Visser, P. M. and Huisman, J. 2014. Contrasting effects of rising $CO_2$ on primary production and ecological stoichiometry at different nutrient levels. *Ecol. Lett.*17, 951–960.

Wand, S. E., Midgley, G. F., Jones, M. C., and Curtis, P. C., 1999. Responses of wild C4 and C3 grass (Poaceae) species to elevated atmospheric $CO_2$ concentration: a meta-analytic test of current theories and perceptions: *Global Change Biology* 5, 723-741.

Warren, R., 2011. The role of interactions in a world implementing adaptation and mitigation solutions to climate change: phil. Trans. R. Soc. A 369, 217-241.

Webster, N. S., Wagner, M. and Negri, A. P. 2018. Microbial conservation in the Anthropocene. *Environ. Microbiol.* 20, 1925–1928.

Weng, Z. H. et al. 2017. Biochar built soil carbon over a decade by stabilizing rhizodeposits. *Nat. Clim. Change*, 7, 371–376.

www.ingramcontent.com/pod-product-compliance
Lightning Source LLC
LaVergne TN
LVHW010719210926
848485LV00008B/1456/J